I0065869

A Comprehensive Handbook of Plant Science

A Comprehensive Handbook of Plant Science

Editor: **Bernard Watts**

R CALLISTO REFERENCE

www.callistoreference.com

Callisto Reference,
118-35 Queens Blvd., Suite 400,
Forest Hills, NY 11375, USA

Visit us on the World Wide Web at:
www.callistoreference.com

© Callisto Reference, 2018

This book contains information obtained from authentic and highly regarded sources. All chapters are published with permission under the Creative Commons Attribution Share Alike License or equivalent. A wide variety of references are listed. Permissions and sources are indicated; for detailed attributions, please refer to the permissions page. Reasonable efforts have been made to publish reliable data and information, but the authors, editors and publisher cannot assume any responsibility for the vailidity of all materials or the consequences of their use.

ISBN: 978-1-64116-000-1 (Hardback)

Trademark Notice: Registered trademark of products or corporate names are used only for explanation and identification without intent to infringe.

Cataloging-in-Publication Data

A comprehensive handbook of plant science / edited by Bernard Watts.
    p. cm.
Includes bibliographical references and index.
ISBN 978-1-64116-000-1
1. Botany. 2. Plants. I. Watts, Bernard.
QK45.2 .C66 2018
580--dc23

Table of Contents

Preface

Plant science or botany refers to the holistic study of plants including plant structure, plant reproduction, evolutionary relationships, plant growth, plant taxonomy, plant diseases, plant differentiation, etc. It is a vast subject, which includes methods like optical microscopy, electron microscopy, live cell imaging, etc. to better understand the evolution and processes of plants. The book studies, analyses and upholds the pillars of plant science and its utmost significance in modern times. Different approaches, evaluations and methodologies have been included in it. Such selected concepts that redefine plant science have been presented in the text. It aims to serve as a resource guide for students and experts alike and contribute to the growth of the discipline.

A foreword of all Chapters of the book is provided below:

Chapter 1 - The branch of biology which studies plants is known as botany. Genomics, proteomics and DNA sequences are new techniques that have helped in developing an exact way of classifying plants. The topics of interest related to the subject are plant growth, structure, reproduction, evolutionary relations and plant taxonomy. The chapter on botany offers an insightful focus, keeping in mind the complex subject matter; **Chapter 2 -** Botany can be divided into various subdisciplines like agronomy, ethnobotany, paleobotany and bryology. Plant morphology studies the external parts of a plant whereas agronomy studies the plants used for the purpose of food, fiber and fuel. This chapter provides a plethora of interdisciplinary topics for better comprehension of botany; **Chapter 3 -** Plants can be classified into different groups such as embryophyte, bryophyte, etc. Embryophytes are the most common type of green plants found on the surface of the Earth. They include mosses, liverworts, ferns, gymnosperms and lycophytes. The major classifications of plants are dealt with great details in the chapter; **Chapter 4 -** Plant physiology studies the functioning of a plant. The processes studied under this subject are plant nutrition, photoperiodism, plant hormone, plant pathology and photomorphogenesis. The section on plant physiology offers an insightful focus, keeping in mind the complex subject matter; **Chapter 5 -** Photosynthesis is the process that is responsible for providing nutrition to plants as well as producing oxygen. The other important aspects related to plant science are phytochemistry, seedling, tropism and flora. This chapter discusses the essential aspects of plant science in a critical manner providing key analysis to the subject matter.

I would like to thank the entire editorial team who made sincere efforts for this book and my family who supported me in my efforts of working on this book. I take this opportunity to thank all those who have been a guiding force throughout my life.

Editor

An Introduction to Botany

The branch of biology which studies plants is known as botany. Genomics, proteomics and DNA sequences are new techniques that have helped in developing an exact way of classifying plants. The topics of interest related to the subject are plant growth, structure, reproduction, evolutionary relations and plant taxonomy. The chapter on botany offers an insightful focus, keeping in mind the complex subject matter.

Botany

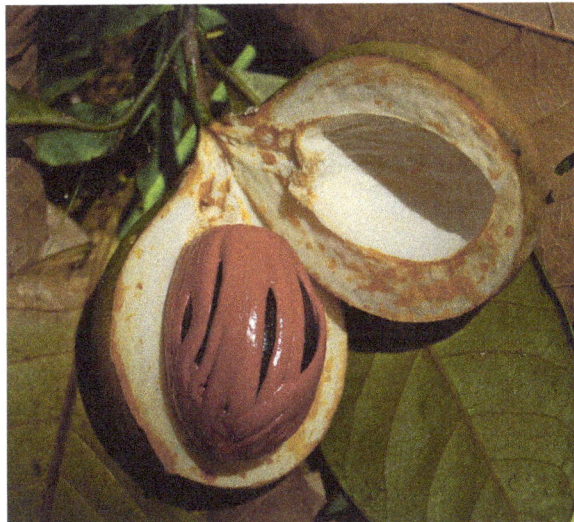

The fruit of *Myristica fragrans*, a species native to Indonesia, is the source of two valuable spices, the red aril (mace) enclosing the dark brown nutmeg

Botany, also called plant science(s), plant biology or phytology, is the science of plant life and a branch of biology. A botanist or plant scientist is a scientist who specialises in this field. The term "botany" comes from the Ancient Greek word *botanē* meaning "pasture", "grass", or "fodder"; is in turn derived from *boskein*, "to feed" or "to graze". Traditionally, botany has also included the study of fungi and algae by mycologists and phycologists respectively, with the study of these three groups of organisms remaining within the sphere of interest of the International Botanical Congress. Nowadays, botanists (in the strict sense) study approximately 410,000 species of land plants of which some 391,000 species are vascular plants (including ca 369,000 species of flowering plants), and ca 20,000 are bryophytes.

Botany originated in prehistory as herbalism with the efforts of early humans to identify – and later cultivate – edible, medicinal and poisonous plants, making it one of the oldest branches of science. Medieval physic gardens, often attached to monasteries, contained plants of medical importance. They were forerunners of the first botanical gardens attached to universities, founded from the 1540s onwards. One of the earliest was the Padua botanical garden. These gardens facilitated the academic study of plants. Efforts to catalogue and describe their collections were the beginnings of plant taxonomy, and led in 1753 to the binomial system of Carl Linnaeus that remains in use to this day.

In the 19th and 20th centuries, new techniques were developed for the study of plants, including methods of optical microscopy and live cell imaging, electron microscopy, analysis of chromosome number, plant chemistry and the structure and function of enzymes and other proteins. In the last two decades of the 20th century, botanists exploited the techniques of molecular genetic analysis, including genomics and proteomics and DNA sequences to classify plants more accurately.

Modern botany is a broad, multidisciplinary subject with inputs from most other areas of science and technology. Research topics include the study of plant structure, growth and differentiation, reproduction, biochemistry and primary metabolism, chemical products, development, diseases, evolutionary relationships, systematics, and plant taxonomy. Dominant themes in 21st century plant science are molecular genetics and epigenetics, which are the mechanisms and control of gene expression during differentiation of plant cells and tissues. Botanical research has diverse applications in providing staple foods, materials such as timber, oil, rubber, fibre and drugs, in modern horticulture, agriculture and forestry, plant propagation, breeding and genetic modification, in the synthesis of chemicals and raw materials for construction and energy production, in environmental management, and the maintenance of biodiversity.

History

Early Botany

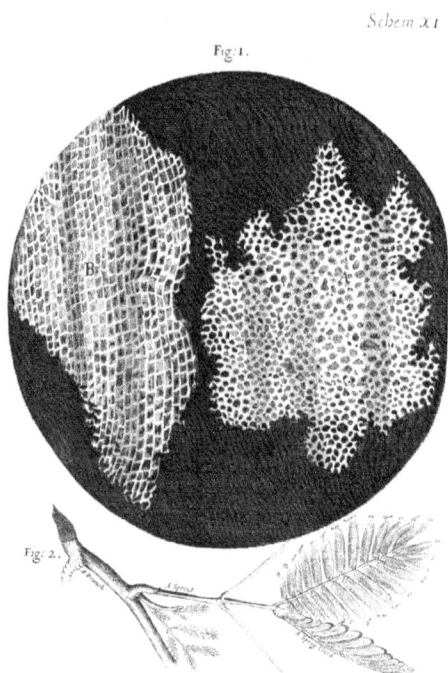

An engraving of the cells of cork, from Robert Hooke's *Micrographia*, 1665

Botany originated as herbalism, the study and use of plants for their medicinal properties. Many records of the Holocene period date early botanical knowledge as far back as 10,000 years ago. This early unrecorded knowledge of plants was discovered in ancient sites of human occupation within Tennessee, which make up much of the Cherokee land today. The early recorded history of botany includes many ancient writings and plant classifications. Examples of early botanical works have been found in ancient texts from India dating back to before 1100 BC, in archaic Avestan writings, and in works from China before it was unified in 221 BC.

Modern botany traces its roots back to Ancient Greece specifically to Theophrastus (c. 371–287 BC), a student of Aristotle who invented and described many of its principles and is widely regarded in the scientific community as the "Father of Botany". His major works, *Enquiry into Plants* and *On the Causes of Plants*, constitute the most important contributions to botanical science until the Middle Ages, almost seventeen centuries later.

Another work from Ancient Greece that made an early impact on botany is *De Materia Medica*, a five-volume encyclopedia about herbal medicine written in the middle of the first century by Greek physician and pharmacologist Pedanius Dioscorides. *De Materia Medica* was widely read for more than 1,500 years. Important contributions from the medieval Muslim world include Ibn Wahshiyya's *Nabatean Agriculture*, Abū Ḥanīfa Dīnawarī's (828–896) the *Book of Plants*, and Ibn Bassal's *The Classification of Soils*. In the early 13th century, Abu al-Abbas al-Nabati, and Ibn al-Baitar (d. 1248) wrote on botany in a systematic and scientific manner.

In the mid-16th century, "botanical gardens" were founded in a number of Italian universities – the Padua botanical garden in 1545 is usually considered to be the first which is still in its original location. These gardens continued the practical value of earlier "physic gardens", often associated with monasteries, in which plants were cultivated for medical use. They supported the growth of botany as an academic subject. Lectures were given about the plants grown in the gardens and their medical uses demonstrated. Botanical gardens came much later to northern Europe; the first in England was the University of Oxford Botanic Garden in 1621. Throughout this period, botany remained firmly subordinate to medicine.

German physician Leonhart Fuchs (1501–1566) was one of "the three German fathers of botany", along with theologian Otto Brunfels (1489–1534) and physician Hieronymus Bock (1498–1554) (also called Hieronymus Tragus). Fuchs and Brunfels broke away from the tradition of copying earlier works to make original observations of their own. Bock created his own system of plant classification.

Physician Valerius Cordus (1515–1544) authored a botanically and pharmacologically important herbal *Historia Plantarum* in 1544 and a pharmacopoeia of lasting importance, the *Dispensatorium* in 1546. Naturalist Conrad von Gesner (1516–1565) and herbalist John Gerard (1545–c. 1611) published herbals covering the medicinal uses of plants. Naturalist Ulisse Aldrovandi (1522–1605) was considered the *father of natural history*, which included the study of plants. In 1665, using an early microscope, Polymath Robert Hooke discovered cells, a term he coined, in cork, and a short time later in living plant tissue.

Early Modern Botany

During the 18th century, systems of plant identification were developed comparable to dichotomous keys, where unidentified plants are placed into taxonomic groups (e.g. family, genus and species) by making a series of choices between pairs of characters. The choice and sequence of the characters may be artificial in keys designed purely for identification (diagnostic keys) or more closely related to the natural or phyletic order of the taxa in synoptic keys. By the 18th century, new plants for study were arriving in Europe in increasing numbers from newly discovered countries and the European colonies worldwide. In 1753 Carl von Linné (Carl Linnaeus) published his Species Plantarum, a hierarchical classification of plant species that remains the reference point for modern botanical nomenclature. This established a standardised binomial or two-part naming scheme where the first name represented the genus and the second identified the species within the genus. For the purposes of identification, Linnaeus's *Systema Sexuale* classified plants into 24 groups according to the number of their male sexual organs. The 24th group, *Cryptogamia*, included all plants with concealed reproductive parts, mosses, liverworts, ferns, algae and fungi.

The Linnaean Garden of Linnaeus' residence in Uppsala, Sweden, was planted according to his *Systema sexuale*

Increasing knowledge of plant anatomy, morphology and life cycles led to the realisation that there were more natural affinities between plants than the artificial sexual system of Linnaeus. Adanson (1763), de Jussieu (1789), and Candolle (1819) all proposed various alternative natural systems of classification that grouped plants using a wider range of shared characters and were widely followed. The Candollean system reflected his ideas of the progression of morphological complexity and the later classification by Bentham and Hooker, which was influential until the mid-19th century, was influenced by Candolle's approach. Darwin's publication of the *Origin of Species* in 1859 and his concept of common descent required modifications to the Candollean system to reflect evolutionary relationships as distinct from mere morphological similarity.

Botany was greatly stimulated by the appearance of the first "modern" textbook, Matthias Schleiden's *Grundzüge der Wissenschaftlichen Botanik*, published in English in 1849 as *Principles of Scientific Botany*. Schleiden was a microscopist and an early plant anatomist who co-founded the cell theory with Theodor Schwann and Rudolf Virchow and was among the first to grasp the significance of the cell nucleus that had been described by Robert Brown in 1831. In 1855, Adolf Fick formulated Fick's laws that enabled the calculation of the rates of molecular diffusion in biological systems.

Echeveria glauca in a Connecticut greenhouse. Botany uses Latin names for identification, here, the specific name *glauca* means blue

Late Modern Botany

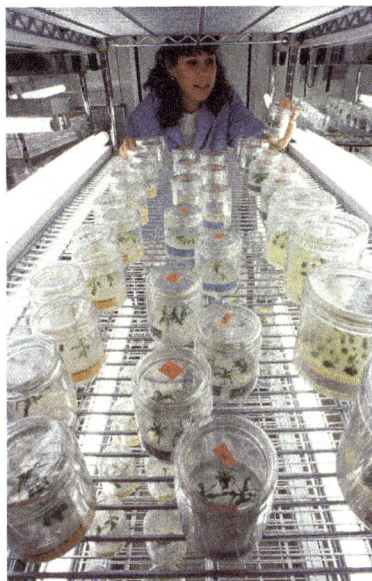

Micropropagation of transgenic plants

Building upon the gene-chromosome theory of heredity that originated with Gregor Mendel (1822–1884), August Weismann (1834–1914) proved that inheritance only takes place through gametes. No other cells can pass on inherited characters. The work of Katherine Esau (1898–1997) on plant anatomy is still a major foundation of modern botany. Her books *Plant Anatomy* and *Anatomy of Seed Plants* have been key plant structural biology texts for more than half a century.

The discipline of plant ecology was pioneered in the late 19th century by botanists such as Eugenius Warming, who produced the hypothesis that plants form communities, and his mentor and successor Christen C. Raunkiær whose system for describing plant life forms is still in use today. The concept that the composition of plant communities such as temperate broadleaf forest changes by a process of ecological succession was developed by Henry Chandler Cowles, Arthur Tansley and Frederic Clements. Clements is credited with the idea of climax vegetation as the most complex vegetation that an environment can support and Tansley introduced the concept of ecosystems to biology. Building on the extensive earlier work of Alphonse de Candolle, Nikolai Vavilov (1887–1943) produced accounts of the biogeography, centres of origin, and evolutionary history of economic plants.

Biologist and statistician Ronald Fisher

Particularly since the mid-1960s there have been advances in understanding of the physics of plant physiological processes such as transpiration (the transport of water within plant tissues), the temperature dependence of rates of water evaporation from the leaf surface and the molecular diffusion of water vapour and carbon dioxide through stomatal apertures. These developments, coupled with new methods for measuring the size of stomatal apertures, and the rate of photosynthesis have enabled precise description of the rates of gas exchange between plants and the atmosphere. Innovations in statistical analysis by Ronald Fisher, Frank Yates and others at Rothamsted Experimental Station facilitated rational experimental design and data analysis in botanical research. The discovery and identification of the auxin plant hormones by Kenneth V. Thimann in 1948 enabled regulation of plant growth by externally applied chemicals. Frederick Campion Steward pioneered techniques of micropropagation and plant tissue culture controlled by plant hormones. The synthetic auxin 2,4-Dichlorophenoxyacetic acid or 2,4-D was one of the first commercial synthetic herbicides.

20th century developments in plant biochemistry have been driven by modern techniques of organic chemical analysis, such as spectroscopy, chromatography and electrophoresis. With the rise of the related molecular-scale biological approaches of molecular biology, genomics, proteomics and metabolomics, the relationship between the plant genome and most aspects of the biochemistry, physiology, morphology and behaviour of plants can be subjected to detailed experimental analysis. The concept originally stated by Gottlieb Haberlandt in 1902 that all plant cells are totipotent and can be grown *in vitro* ultimately enabled the use of genetic engineering experimentally to knock out a gene or genes responsible for a specific trait, or to add genes such as GFP that report when a gene of interest is being expressed. These technologies enable the biotechnological use of whole plants or plant cell cultures grown in bioreactors to synthesise pesticides, antibiotics or other pharmaceuticals, as well as the practical application of genetically modified crops designed for traits such as improved yield.

Modern morphology recognises a continuum between the major morphological categories of root, stem (caulome), leaf (phyllome) and trichome. Furthermore, it emphasises structural dynamics.

Modern systematics aims to reflect and discover phylogenetic relationships between plants. Modern Molecular phylogenetics largely ignores morphological characters, relying on DNA sequences as data. Molecular analysis of DNA sequences from most families of flowering plants enabled the Angiosperm Phylogeny Group to publish in 1998 a phylogeny of flowering plants, answering many of the questions about relationships among angiosperm families and species. The theoretical possibility of a practical method for identification of plant species and commercial varieties by DNA barcoding is the subject of active current research.

Scope and Importance

Botany involves the recording and description of plants, such as this herbarium specimen of the lady fern *Athyrium filix-femina*

The study of plants is vital because they underpin almost all animal life on Earth by generating a large proportion of the oxygen and food that provide humans and other organisms with aerobic respiration with the chemical energy they need to exist. Plants, algae and cyanobacteria are the major groups of organisms that carry out photosynthesis, a process that uses the energy of sunlight to convert water and carbon dioxide into sugars that can be used both as a source of chemical energy and of organic molecules that are used in the structural components of cells. As a by-product of photosynthesis, plants release oxygen into the atmosphere, a gas that is required by nearly all living things to carry out cellular respiration. In addition, they are influential in the global carbon and water cycles and plant roots bind and stabilise soils, preventing soil erosion. Plants are crucial to the future of human society as they provide food, oxygen, medicine, and products for people, as well as creating and preserving soil.

Historically, all living things were classified as either animals or plants and botany covered the study of all organisms not considered animals. Botanists examine both the internal functions and processes within plant organelles, cells, tissues, whole plants, plant populations and plant communities. At each of these levels, a botanist may be concerned with the classification (taxonomy), phylogeny and evolution, structure (anatomy and morphology), or function (physiology) of plant life.

The strictest definition of "plant" includes only the "land plants" or embryophytes, which include seed plants (gymnosperms, including the pines, and flowering plants) and the free-sporing cryptogams including ferns, clubmosses, liverworts, hornworts and mosses. Embryophytes are multicellular eukaryotes descended from an ancestor that obtained its energy from sunlight by photosynthesis. They have life cycles with alternating haploid and diploid phases. The sexual haploid phase of embryophytes, known as the gametophyte, nurtures the developing diploid embryo sporophyte within its tissues for at least part of its life, even in the seed plants, where the gametophyte itself is nurtured by its parent sporophyte. Other groups of organisms that were previously studied by botanists include bacteria (now studied in bacteriology), fungi (mycology) – including lichen-forming fungi (lichenology), non-chlorophyte algae (phycology), and viruses (virology). However, attention is still given to these groups by botanists, and fungi (including lichens) and photosynthetic protists are usually covered in introductory botany courses.

Palaeobotanists study ancient plants in the fossil record to provide information about the evolutionary history of plants. Cyanobacteria, the first oxygen-releasing photosynthetic organisms on Earth, are thought to have given rise to the ancestor of plants by entering into an endosymbiotic relationship with an early eukaryote, ultimately becoming the chloroplasts in plant cells. The new photosynthetic plants (along with their algal relatives) accelerated the rise in atmospheric oxygen started by the cyanobacteria, changing the ancient oxygen-free, reducing, atmosphere to one in which free oxygen has been abundant for more than 2 billion years.

Among the important botanical questions of the 21st century are the role of plants as primary producers in the global cycling of life's basic ingredients: energy, carbon, oxygen, nitrogen and water, and ways that our plant stewardship can help address the global environmental issues of resource management, conservation, human food security, biologically invasive organisms, carbon sequestration, climate change, and sustainability.

Human Nutrition

The food we eat comes directly or indirectly from plants such as rice

Virtually all staple foods come either directly from primary production by plants, or indirectly from animals that eat them. Plants and other photosynthetic organisms are at the base of most food chains because they use the energy from the sun and nutrients from the soil and atmosphere, converting them into a form that can be used by animals. This is what ecologists call the first trophic level. The modern forms of the major staple foods, such as maize, rice, wheat and other cereal grasses, pulses, bananas and plantains, as well as flax and cotton grown for their fibres, are the outcome of prehistoric selection over thousands of years from among wild ancestral plants with the most desirable characteristics.

Botanists study how plants produce food and how to increase yields, for example through plant breeding, making their work important to mankind's ability to feed the world and provide food security for future generations. Botanists also study weeds, which are a considerable problem in agriculture, and the biology and control of plant pathogens in agriculture and natural ecosystems. Ethnobotany is the study of the relationships between plants and people. When applied to the investigation of historical plant–people relationships ethnobotany may be referred to as archaeobotany or palaeoethnobotany. Some of the earliest plant-people relationships arose between the indigenous people of Canada in identifying edible plants from inedible plants. This relationship the indigenous people had with plants was recorded by ethnobotanists.

Plant Biochemistry

Plant biochemistry is the study of the chemical processes used by plants. Some of these processes are used in their primary metabolism like the photosynthetic Calvin cycle and crassulacean acid metabolism. Others make specialised materials like the cellulose and lignin used to build their bodies, and secondary products like resins and aroma compounds.

Plants make various photosynthetic pigments, some of which can be seen here through paper chromatography.

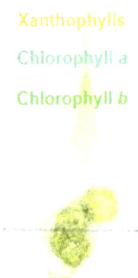

Xanthophylls

Chlorophyll a

Chlorophyll b

Plants and various other groups of photosynthetic eukaryotes collectively known as "algae" have unique organelles known as chloroplasts. Chloroplasts are thought to be descended from cyanobacteria that formed endosymbiotic relationships with ancient plant and algal ancestors. Chloroplasts and cyanobacteria contain the blue-green pigment chlorophyll a. Chlorophyll a (as well as its plant and green algal-specific cousin chlorophyll b)[a] absorbs light in the blue-violet and orange/red parts of the spectrum while reflecting and transmitting the green light that we see as the characteristic colour of these organisms. The energy in the red and blue light that these pigments absorb is used by chloroplasts to make energy-rich carbon compounds from carbon dioxide and water by oxygenic photosynthesis, a process that generates molecular oxygen (O_2) as a by-product.

The Calvin cycle *(Interactive diagram)* The Calvin cycle incorporates carbon dioxide into sugar molecules.

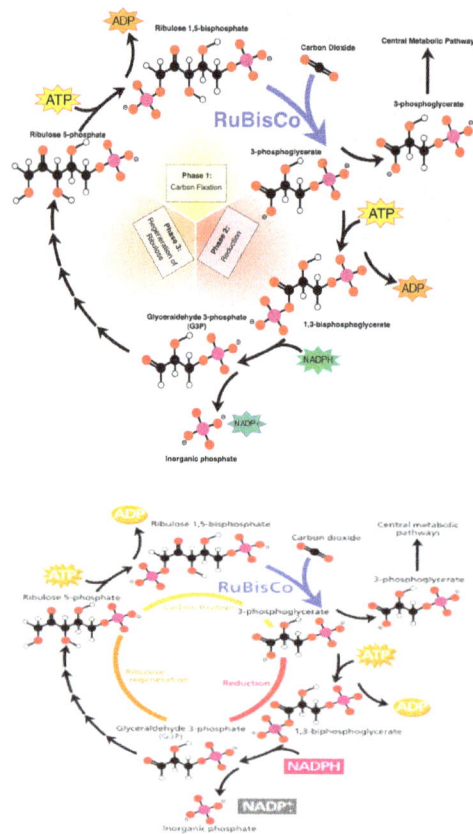

The light energy captured by chlorophyll *a* is initially in the form of electrons (and later a proton gradient) that's used to make molecules of ATP and NADPH which temporarily store and transport energy. Their energy is used in the light-independent reactions of the Calvin cycle by the enzyme rubisco to produce molecules of the 3-carbon sugar glyceraldehyde 3-phosphate (G3P). Glyceraldehyde 3-phosphate is the first product of photosynthesis and the raw material from which glucose and almost all other organic molecules of biological origin are synthesised. Some of the glucose is converted to starch which is stored in the chloroplast. Starch is the characteristic energy store of most land plants and algae, while inulin, a polymer of fructose is used for the same purpose in the sunflower family Asteraceae. Some of the glucose is converted to sucrose (common table sugar) for export to the rest of the plant.

Unlike in animals (which lack chloroplasts), plants and their eukaryote relatives have delegated many biochemical roles to their chloroplasts, including synthesising all their fatty acids, and most amino acids. The fatty acids that chloroplasts make are used for many things, such as providing material to build cell membranes out of and making the polymer cutin which is found in the plant cuticle that protects land plants from drying out.

Plants synthesise a number of unique polymers like the polysaccharide molecules cellulose, pectin and xyloglucan from which the land plant cell wall is constructed. Vascular land plants make lignin, a polymer used to strengthen the secondary cell walls of xylem tracheids and vessels to keep them from collapsing when a plant sucks water through them under water stress. Lignin is also used in other cell types like sclerenchyma fibres that provide structural support for a plant and is a major constituent of wood. Sporopollenin is a chemically resistant polymer found in the outer cell walls of spores and pollen of land plants responsible for the survival of early land plant spores and the pollen of seed plants in the fossil record. It is widely regarded as a marker for the start of land plant evolution during the Ordovician period. The concentration of carbon dioxide in the atmosphere today is much lower than it was when plants emerged onto land during the Ordovician and Silurian periods. Many monocots like maize and the pineapple and some dicots like the Asteraceae have since independently evolved pathways like Crassulacean acid metabolism and the C_4 carbon fixation pathway for photosynthesis which avoid the losses resulting from photorespiration in the more common C_3 carbon fixation pathway. These biochemical strategies are unique to land plants.

Medicine and Materials

Tapping a rubber tree in Thailand

Phytochemistry is a branch of plant biochemistry primarily concerned with the chemical substances produced by plants during secondary metabolism. Some of these compounds are toxins such as the alkaloid coniine from hemlock. Others, such as the essential oils peppermint oil and lemon oil are useful for their aroma, as flavourings and spices (e.g., capsaicin), and in medicine as pharmaceuticals as in opium from opium poppies. Many medicinal and recreational drugs, such as tetrahydrocannabinol (active ingredient in cannabis), caffeine, morphine and nicotine come directly from plants. Others are simple derivatives of botanical natural products. For example, the pain killer aspirin is the acetyl ester of salicylic acid, originally isolated from the bark of willow trees, and a wide range of opiate painkillers like heroin are obtained by chemical modification of morphine obtained from the opium poppy. Popular stimulants come from plants, such as caffeine from coffee, tea and chocolate, and nicotine from tobacco. Most alcoholic beverages come from fermentation of carbohydrate-rich plant products such as barley (beer), rice (sake) and grapes (wine). Native Americans have used various plants as ways of treating illness or disease for thousands of years. This knowledge Native Americans have on plants has been recorded by enthnobotanists and then in turn has been used by pharmaceutical companies as a way of drug discovery.

Plants can synthesise useful coloured dyes and pigments such as the anthocyanins responsible for the red colour of red wine, yellow weld and blue woad used together to produce Lincoln green, indoxyl, source of the blue dye indigo traditionally used to dye denim and the artist's pigments gamboge and rose madder. Sugar, starch, cotton, linen, hemp, some types of rope, wood and particle boards, papyrus and paper, vegetable oils, wax, and natural rubber are examples of commercially important materials made from plant tissues or their secondary products. Charcoal, a pure form of carbon made by pyrolysis of wood, has a long history as a metal-smelting fuel, as a filter material and adsorbent and as an artist's material and is one of the three ingredients of gunpowder. Cellulose, the world's most abundant organic polymer, can be converted into energy, fuels, materials and chemical feedstock. Products made from cellulose include rayon and cellophane, wallpaper paste, biobutanol and gun cotton. Sugarcane, rapeseed and soy are some of the plants with a highly fermentable sugar or oil content that are used as sources of biofuels, important alternatives to fossil fuels, such as biodiesel. Sweetgrass was used by NativeAmericanse to ward of bugs like mosquitoes. These bug repelling properties of sweetgrass were later found by the American Chemical Society in the molecules phytol and coumarin.

Plant Ecology

Holdridge life zones model relationships between vegetation type, moisture availability and temperature.

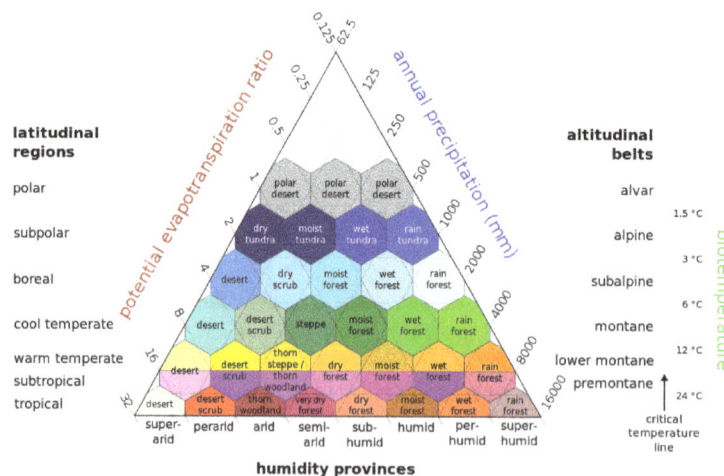

Plant ecology is the science of the functional relationships between plants and their habitats—the environments where they complete their life cycles. Plant ecologists study the composition of local and regional floras, their biodiversity, genetic diversity and fitness, the adaptation of plants to their environment, and their competitive or mutualistic interactions with other species. Some ecologists even rely on empirical data from indigenous people that is gathered by ethnobotanists. This information can relay a great deal of information on how the land once was thousands of years ago and how it has changed over that time. The goals of plant ecology are to understand the causes of their distribution patterns, productivity, environmental impact, evolution, and responses to environmental change.

Plants depend on certain edaphic (soil) and climatic factors in their environment but can modify these factors too. For example, they can change their environment's albedo, increase runoff interception, stabilise mineral soils and develop their organic content, and affect local temperature. Plants compete with other organisms in their ecosystem for resources. They interact with their neighbours at a variety of spatial scales in groups, populations and communities that collectively constitute vegetation. Regions with characteristic vegetation types and dominant plants as well as similar abiotic and biotic factors, climate, and geography make up biomes like tundra or tropical rainforest.

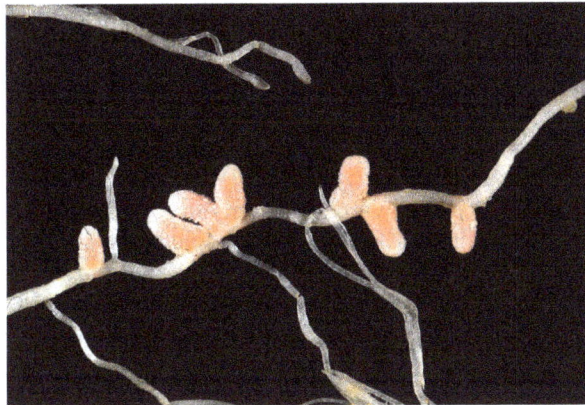

The nodules of *Medicago italica* contain the nitrogen fixing bacterium *Sinorhizobium meliloti*.
The plant provides the bacteria with nutrients and an anaerobic environment,
and the bacteria fix nitrogen for the plant

Herbivores eat plants, but plants can defend themselves and some species are parasitic or even carnivorous. Other organisms form mutually beneficial relationships with plants. For example, mycorrhizal fungi and rhizobia provide plants with nutrients in exchange for food, ants are recruited by ant plants to provide protection, honey bees, bats and other animals pollinate flowers and humans and other animals act as dispersal vectors to spread spores and seeds.

Plants, Climate and Environmental Change

Plant responses to climate and other environmental changes can inform our understanding of how these changes affect ecosystem function and productivity. For example, plant phenology can be a useful proxy for temperature in historical climatology, and the biological impact of climate change and global warming. Palynology, the analysis of fossil pollen deposits in sediments from thousands or millions of years ago allows the reconstruction of past climates. Estimates of atmospheric CO_2 concentrations since the Palaeozoic have been obtained from stomatal densities and the leaf shapes and sizes of ancient land plants. Ozone depletion can expose plants to higher levels of ultraviolet radiation-B (UV-B), resulting in lower growth rates. Moreover, information from studies of community ecology, plant systematics, and taxonomy is essential to understanding vegetation change, habitat destruction and species extinction.

Genetics

A Punnett square depicting a cross between two pea plants heterozygous
for purple (B) and white (b) blossoms

Inheritance in plants follows the same fundamental principles of genetics as in other multicellular organisms. Gregor Mendel discovered the genetic laws of inheritance by studying inherited traits such as shape in *Pisum sativum* (peas). What Mendel learned from studying plants has had far reaching benefits outside of botany. Similarly, "jumping genes" were discovered by Barbara McClintock while she was studying maize. Nevertheless, there are some distinctive genetic differences between plants and other organisms.

Species boundaries in plants may be weaker than in animals, and cross species hybrids are often possible. A familiar example is peppermint, *Mentha × piperita*, a sterile hybrid between *Mentha aquatica* and spearmint, *Mentha spicata*. The many cultivated varieties of wheat are the result of multiple inter- and intra-specific crosses between wild species and their hybrids. Angiosperms with monoecious flowers often have self-incompatibility mechanisms that operate between the pollen and stigma so that the pollen either fails to reach the stigma or fails to germinate and produce male gametes. This is one of several methods used by plants to promote outcrossing. In many land plants the male and female gametes are produced by separate individuals. These species are said to be dioecious when referring to vascular plant sporophytes and dioicous when referring to bryophyte gametophytes.

Unlike in higher animals, where parthenogenesis is rare, asexual reproduction may occur in plants by several different mechanisms. The formation of stem tubers in potato is one example. Particularly in arctic or alpine habitats, where opportunities for fertilisation of flowers by animals are rare, plantlets or bulbs, may develop instead of flowers, replacing sexual reproduction with asexual reproduction and giving rise to clonal populations genetically identical to the parent. This is one of several types of apomixis that occur in plants. Apomixis can also happen in a seed, producing a seed that contains an embryo genetically identical to the parent.

Most sexually reproducing organisms are diploid, with paired chromosomes, but doubling of their chromosome number may occur due to errors in cytokinesis. This can occur early in development to produce an autopolyploid or partly autopolyploid organism, or during normal processes of cellular differentiation to produce some cell types that are polyploid (endopolyploidy), or during gamete formation. An allopolyploid plant may result from a hybridisation event between two different species. Both autopolyploid and allopolyploid plants can often reproduce normally, but may be unable to cross-breed successfully with the parent population because there is a mismatch in chromosome numbers. These plants that are reproductively isolated from the parent species but live within the same geographical area, may be sufficiently successful to form a new species. Some otherwise sterile plant polyploids can still reproduce vegetatively or by seed apomixis, forming clonal populations of identical individuals. Durum wheat is a fertile tetraploid allopolyploid, while bread wheat is a fertile hexaploid. The commercial banana is an example of a sterile, seedless triploid hybrid. Common dandelion is a triploid that produces viable seeds by apomictic seed.

As in other eukaryotes, the inheritance of endosymbiotic organelles like mitochondria and chloroplasts in plants is non-Mendelian. Chloroplasts are inherited through the male parent in gymnosperms but often through the female parent in flowering plants.

Molecular Genetics

Thale cress, *Arabidopsis thaliana*, the first plant to have its genome sequenced, remains the most important model organism

A considerable amount of new knowledge about plant function comes from studies of the molecular genetics of model plants such as the Thale cress, *Arabidopsis thaliana*, a weedy species in the mustard family (Brassicaceae). The genome or hereditary information contained in the genes of this species is encoded by about 135 million base pairs of DNA, forming one of the smallest genomes among flowering plants. *Arabidopsis* was the first plant to have its genome sequenced, in 2000. The sequencing of some other relatively small genomes, of rice (*Oryza sativa*) and *Brachypodium distachyon*, has made them important model species for understanding the genetics, cellular and molecular biology of cereals, grasses and monocots generally.

Model plants such as *Arabidopsis thaliana* are used for studying the molecular biology of plant cells and the chloroplast. Ideally, these organisms have small genomes that are well known or completely sequenced, small stature and short generation times. Corn has been used to study mechanisms of photosynthesis and phloem loading of sugar in C_4 plants. The single celled green alga *Chlamydomonas reinhardtii*, while not an embryophyte itself, contains a green-pigmented chloroplast related to that of land plants, making it useful for study. A red alga *Cyanidioschyzon merolae* has also been used to study some basic chloroplast functions. Spinach, peas, soybeans and a moss *Physcomitrella patens* are commonly used to study plant cell biology.

Agrobacterium tumefaciens, a soil rhizosphere bacterium, can attach to plant cells and infect them with a callus-inducing Ti plasmid by horizontal gene transfer, causing a callus infection called crown gall disease. Schell and Van Montagu (1977) hypothesised that the Ti plasmid could be a natural vector for introducing the Nif gene responsible for nitrogen fixation in the root nodules of legumes and other plant species. Today, genetic modification of the Ti plasmid is one of the main techniques for introduction of transgenes to plants and the creation of genetically modified crops.

Epigenetics

Epigenetics is the study of heritable changes in gene function that cannot be explained by changes in the underlying DNA sequence but cause the organism's genes to behave (or "express themselves") differently. One example of epigenetic change is the marking of the genes by DNA methylation which determines whether they will be expressed or not. Gene expression can also be controlled by repressor proteins that attach to silencer regions of the DNA and prevent that region of the DNA code from being expressed. Epigenetic marks may be added or removed from the DNA during programmed stages of development of the plant, and are responsible, for example, for the differences between anthers, petals and normal leaves, despite the fact that they all have the same underlying genetic code. Epigenetic changes may be temporary or may remain through successive cell divisions for the remainder of the cell's life. Some epigenetic changes have been shown to be heritable, while others are reset in the germ cells.

Epigenetic changes in eukaryotic biology serve to regulate the process of cellular differentiation. During morphogenesis, totipotent stem cells become the various pluripotent cell lines of the embryo, which in turn become fully differentiated cells. A single fertilised egg cell, the zygote, gives rise to the many different plant cell types including parenchyma, xylem vessel elements, phloem sieve tubes, guard cells of the epidermis, etc. as it continues to divide. The process results from the epigenetic activation of some genes and inhibition of others.

Unlike animals, many plant cells, particularly those of the parenchyma, do not terminally differentiate, remaining totipotent with the ability to give rise to a new individual plant. Exceptions include highly lignified cells, the sclerenchyma and xylem which are dead at maturity, and the phloem sieve tubes which lack nuclei. While plants use many of the same epigenetic mechanisms as animals, such as chromatin remodelling, an alternative hypothesis is that plants set their gene expression patterns using positional information from the environment and surrounding cells to determine their developmental fate.

Plant Evolution

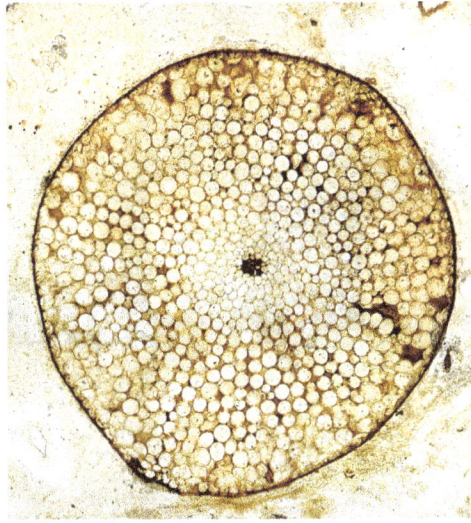

Transverse section of a fossil stem of the Devonian vascular plant *Rhynia gwynne-vaughani*

The chloroplasts of plants have a number of biochemical, structural and genetic similarities to cyanobacteria, (commonly but incorrectly known as "blue-green algae") and are thought to be derived from an ancient endosymbiotic relationship between an ancestral eukaryotic cell and a cyanobacterial resident.

The algae are a polyphyletic group and are placed in various divisions, some more closely related to plants than others. There are many differences between them in features such as cell wall composition, biochemistry, pigmentation, chloroplast structure and nutrient reserves. The algal division Charophyta, sister to the green algal division Chlorophyta, is considered to contain the ancestor of true plants. The Charophyte class Charophyceae and the land plant sub-kingdom Embryophyta together form the monophyletic group or clade Streptophytina.

Nonvascular land plants are embryophytes that lack the vascular tissues xylem and phloem. They include mosses, liverworts and hornworts. Pteridophytic vascular plants with true xylem and phloem that reproduced by spores germinating into free-living gametophytes evolved during the Silurian period and diversified into several lineages during the late Silurian and early Devonian. Representatives of the lycopods have survived to the present day. By the end of the Devonian period, several groups, including the lycopods, sphenophylls and progymnosperms, had independently evolved "megaspory" – their spores were of two distinct sizes, larger megaspores and smaller microspores. Their reduced gametophytes developed from megaspores retained within the spore-producing organs (megasporangia) of the sporophyte, a condition known as endospory. Seeds consist of an endosporic megasporangium surrounded by one or two sheathing layers (integuments). The young sporophyte develops within the seed, which on germination splits to release it. The earliest known seed plants date from the latest Devonian Famennian stage. Following the evolution of the seed habit, seed plants diversified, giving rise to a number of now-extinct groups, including seed ferns, as well as the modern gymnosperms and angiosperms. Gymnosperms produce "naked seeds" not fully enclosed in an ovary; modern representatives include conifers, cycads, *Ginkgo*, and Gnetales. Angiosperms produce seeds enclosed in a structure such as a carpel or an ovary. Ongoing research on the molecular phylogenetics of living plants appears to show that the angiosperms are a sister clade to the gymnosperms.

Plant Physiology

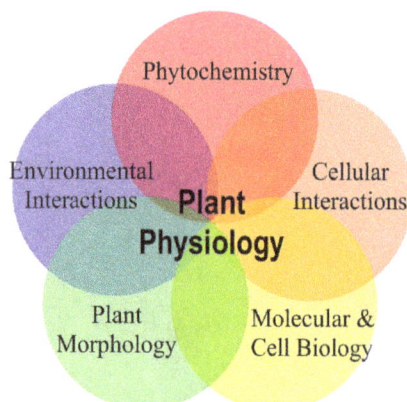

Five of the key areas of study within plant physiology

Plant physiology encompasses all the internal chemical and physical activities of plants associated with life. Chemicals obtained from the air, soil and water form the basis of all plant metabolism. The energy of sunlight, captured by oxygenic photosynthesis and released by cellular respiration, is the basis of almost all life. Photoautotrophs, including all green plants, algae and cyanobacteria gather energy directly from sunlight by photosynthesis. Heterotrophs including all animals, all fungi, all completely parasitic plants, and non-photosynthetic bacteria take in organic molecules produced by photoautotrophs and respire them or use them in the construction of cells and tissues. Respiration is the oxidation of carbon compounds by breaking them down into simpler structures to release the energy they contain, essentially the opposite of photosynthesis.

Molecules are moved within plants by transport processes that operate at a variety of spatial scales. Subcellular transport of ions, electrons and molecules such as water and enzymes occurs across cell membranes. Minerals and water are transported from roots to other parts of the plant in the transpiration stream. Diffusion, osmosis, and active transport and mass flow are all different ways transport can occur. Examples of elements that plants need to transport are nitrogen, phosphorus, potassium, calcium, magnesium, and sulphur. In vascular plants, these elements are extracted from the soil as soluble ions by the roots and transported throughout the plant in the xylem. Most of the elements required for plant nutrition come from the chemical breakdown of soil minerals. Sucrose produced by photosynthesis is transported from the leaves to other parts of the plant in the phloem and plant hormones are transported by a variety of processes.

Plant Hormones

1 An oat coleoptile with the sun overhead. Auxin (pink) is evenly distributed in its tip.

2 With the sun at an angle and only shining on one side of the shoot, auxin moves to the opposite side and stimulates cell elongation there.

3 and **4** Extra growth on that side causes the shoot to bend towards the sun.

Plants are not passive, but respond to external signals such as light, touch, and injury by moving or growing towards or away from the stimulus, as appropriate. Tangible evidence of touch sensitivity is the almost instantaneous collapse of leaflets of *Mimosa pudica*, the insect traps of Venus flytrap and bladderworts, and the pollinia of orchids.

The hypothesis that plant growth and development is coordinated by plant hormones or plant growth regulators first emerged in the late 19th century. Darwin experimented on the movements of plant shoots and roots towards light and gravity, and concluded "It is hardly an exaggeration to say that the tip of the radicle .. acts like the brain of one of the lower animals .. directing the several movements". About the same time, the role of auxins (from the Greek auxein, to grow) in control of plant growth was first outlined by the Dutch scientist Frits Went. The first known auxin, indole-3-acetic acid (IAA), which promotes cell growth, was only isolated from plants about 50 years later. This compound mediates the tropic responses of shoots and roots towards light and gravity. The finding in 1939 that plant callus could be maintained in culture containing IAA, followed by the observation in 1947 that it could be induced to form roots and shoots by controlling the concentration of growth hormones were key steps in the development of plant biotechnology and genetic modification.

Venus's fly trap, *Dionaea muscipula*, showing the touch-sensitive insect trap in action

Cytokinins are a class of plant hormones named for their control of cell division or cytokinesis. The natural cytokinin zeatin was discovered in corn, *Zea mays*, and is a derivative of the purine adenine. Zeatin is produced in roots and transported to shoots in the xylem where it promotes cell division, bud development, and the greening of chloroplasts. The gibberelins, such as Gibberelic acid are diterpenes synthesised from acetyl CoA via the mevalonate pathway. They are involved in the promotion of germination and dormancy-breaking in seeds, in regulation of plant height by controlling stem elongation and the control of flowering. Abscisic acid (ABA) occurs in all land plants except liverworts, and is synthesised from carotenoids in the chloroplasts and other plastids. It inhibits cell division, promotes seed maturation, and dormancy, and promotes stomatal closure. It was so named because it was originally thought to control abscission. Ethylene is a gaseous hormone that is produced in all higher plant tissues from methionine. It is now known to be the hormone that stimulates or regulates fruit ripening and abscission, and it, or the synthetic growth regulator ethephon which is rapidly metabolised to produce ethylene, are used on industrial scale to promote ripening of cotton, pineapples and other climacteric crops.

Another class of phytohormones is the jasmonates, first isolated from the oil of *Jasminum grandiflorum* which regulates wound responses in plants by unblocking the expression of genes required in the systemic acquired resistance response to pathogen attack.

In addition to being the primary energy source for plants, light functions as a signalling device, providing information to the plant, such as how much sunlight the plant receives each day. This can result in adaptive changes in a process known as photomorphogenesis. Phytochromes are the photoreceptors in a plant that are sensitive to light.

Plant Anatomy and Morphology

A nineteenth-century illustration showing the morphology of the roots, stems, leaves and flowers of the rice plant *Oryza sativa*

Plant anatomy is the study of the structure of plant cells and tissues, whereas plant morphology is the study of their external form. All plants are multicellular eukaryotes, their DNA stored in nuclei. The characteristic features of plant cells that distinguish them from those of animals and fungi include a primary cell wall composed of the polysaccharides cellulose, hemicellulose and pectin, larger vacuoles than in animal cells and the presence of plastids with unique photosynthetic and biosynthetic functions as in the chloroplasts. Other plastids contain storage products such as starch (amyloplasts) or lipids (elaioplasts). Uniquely, streptophyte cells and those of the green algal order Trentepohliales divide by construction of a phragmoplast as a template for building a cell plate late in cell division.

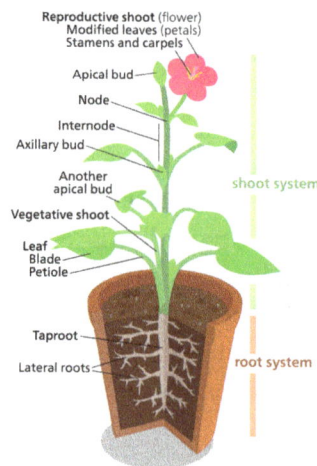

A diagram of a "typical" eudicot, the most common type of plant (three-fifths of all plant species). No plant actually looks exactly like this though.

The bodies of vascular plants including clubmosses, ferns and seed plants (gymnosperms and angiosperms) generally have aerial and subterranean subsystems. The shoots consist of stems bearing green photosynthesising leaves and reproductive structures. The underground vascularised roots bear root hairs at their tips and generally lack chlorophyll. Non-vascular plants, the liverworts, hornworts and mosses do not produce ground-penetrating vascular roots and most of the plant participates in photosynthesis. The sporophyte generation is nonphotosynthetic in liverworts but may be able to contribute part of its energy needs by photosynthesis in mosses and hornworts.

The root system and the shoot system are interdependent – the usually nonphotosynthetic root system depends on the shoot system for food, and the usually photosynthetic shoot system depends on water and minerals from the root system. Cells in each system are capable of creating cells of the other and producing adventitious shoots or roots. Stolons and tubers are examples of shoots that can grow roots. Roots that spread out close to the surface, such as those of willows, can produce shoots and ultimately new plants. In the event that one of the systems is lost, the other can often regrow it. In fact it is possible to grow an entire plant from a single leaf, as is the case with *Saintpaulia*, or even a single cell – which can dedifferentiate into a callus (a mass of unspecialised cells) that can grow into a new plant. In vascular plants, the xylem and phloem are the conductive tissues that transport resources between shoots and roots. Roots are often adapted to store food such as sugars or starch, as in sugar beets and carrots.

Stems mainly provide support to the leaves and reproductive structures, but can store water in succulent plants such as cacti, food as in potato tubers, or reproduce vegetatively as in the stolons of strawberry plants or in the process of layering. Leaves gather sunlight and carry out photosynthesis. Large, flat, flexible, green leaves are called foliage leaves. Gymnosperms, such as conifers, cycads, *Ginkgo*, and gnetophytes are seed-producing plants with open seeds. Angiosperms are seed-producing plants that produce flowers and have enclosed seeds. Woody plants, such as azaleas and oaks, undergo a secondary growth phase resulting in two additional types of tissues: wood (secondary xylem) and bark (secondary phloem and cork). All gymnosperms and many angiosperms are woody plants. Some plants reproduce sexually, some asexually, and some via both means.

Although reference to major morphological categories such as root, stem, leaf, and trichome are useful, one has to keep in mind that these categories are linked through intermediate forms so that a continuum between the categories results. Furthermore, structures can be seen as processes, that is, process combinations.

Systematic Botany

Systematic botany is part of systematic biology, which is concerned with the range and diversity of organisms and their relationships, particularly as determined by their evolutionary history. It involves, or is related to, biological classification, scientific taxonomy and phylogenetics. Biological classification is the method by which botanists group organisms into categories such as genera or species. Biological classification is a form of scientific taxonomy. Modern taxonomy is rooted in the work of Carl Linnaeus, who grouped species according to shared physical characteristics. These groupings have since been revised to align better with the Darwinian principle of common descent – grouping organisms by ancestry rather than superficial characteristics. While scientists do not always agree on how to classify organisms, molecular phylogenetics, which uses DNA sequences as data, has driven many recent revisions along evolutionary lines and is likely to continue to do so. The dominant classification system is called Linnaean taxonomy. It includes ranks and binomial nomenclature. The nomenclature of botanical organisms is codified in the International Code of Nomenclature for algae, fungi, and plants (ICN) and administered by the International Botanical Congress.

A botanist preparing a plant specimen for mounting in the herbarium

Kingdom Plantae belongs to Domain Eukarya and is broken down recursively until each species is separately classified. The order is: Kingdom; Phylum (or Division); Class; Order; Family; Genus (plural *genera*); Species. The scientific name of a plant represents its genus and its species within the genus, resulting in a single worldwide name for each organism. For example, the tiger lily is *Lilium columbianum*. *Lilium* is the genus, and *columbianum* the specific epithet. The combination is the name of the species. When writing the scientific name of an organism, it is proper to capitalise the first letter in the genus and put all of the specific epithet in lowercase.

The evolutionary relationships and heredity of a group of organisms is called its phylogeny. Phylogenetic studies attempt to discover phylogenies. The basic approach is to use similarities based on shared inheritance to determine relationships. As an example, species of *Pereskia* are trees or bushes with prominent leaves. They do not obviously resemble a typical leafless cactus such as an *Echinocactus*. However, both *Pereskia* and *Echinocactus* have spines produced from areoles (highly specialised pad-like structures) suggesting that the two genera are indeed related.

Two Cacti of Very Different Appearance

Pereskia aculeata

Echinocactus grusonii

Although *Pereskia* is a tree with leaves, it has spines and areoles like a more typical cactus, such as *Echinocactus*.

Judging relationships based on shared characters requires care, since plants may resemble one another through convergent evolution in which characters have arisen independently. Some euphorbias have leafless, rounded bodies adapted to water conservation similar to those of globular cacti, but characters such as the structure of their flowers make it clear that the two groups are not closely related. The cladistic method takes a systematic approach to characters, distinguishing between those that carry no information about shared evolutionary history – such as those evolved separately in different groups (homoplasies) or those left over from ancestors (plesiomorphies) – and derived characters, which have been passed down from innovations in a shared ancestor (apomorphies). Only derived characters, such as the spine-producing areoles of cacti, provide evidence for descent from a common ancestor. The results of cladistic analyses are expressed as cladograms: tree-like diagrams showing the pattern of evolutionary branching and descent.

From the 1990s onwards, the predominant approach to constructing phylogenies for living plants has been molecular phylogenetics, which uses molecular characters, particularly DNA sequences, rather than morphological characters like the presence or absence of spines and areoles. The difference is that the genetic code itself is used to decide evolutionary relationships, instead of being used indirectly via the characters it gives rise to. Clive Stace describes this as having "direct access to the genetic basis of evolution." As a simple example, prior to the use of genetic evidence, fungi were thought either to be plants or to be more closely related to plants than animals. Genetic evidence suggests that the true evolutionary relationship of multicelled organisms is as shown in the cladogram below – fungi are more closely related to animals than to plants.

In 1998 the Angiosperm Phylogeny Group published a phylogeny for flowering plants based on an analysis of DNA sequences from most families of flowering plants. As a result of this work, many questions, such as which families represent the earliest branches of angiosperms, have now been answered. Investigating how plant species are related to each other allows botanists to better understand the process of evolution in plants. Despite the study of model plants and increasing use of DNA evidence, there is ongoing work and discussion among taxonomists about how best to classify plants into various taxa. Technological developments such as computers and electron microscopes have greatly increased the level of detail studied and speed at which data can be analysed.

History of Botany

Some traditional tools of botanical science

The history of botany examines the human effort to understand life on Earth by tracing the historical development of the discipline of botany—that part of natural science dealing with organisms traditionally treated as plants.

Rudimentary botanical science began with empirically-based plant lore passed from generation to generation in the oral traditions of paleolithic hunter-gatherers. The first written records of plants were made in the Neolithic Revolution about 10,000 years ago as writing was developed in the settled agricultural communities where plants and animals were first domesticated. The first writings that show human curiosity about plants themselves, rather than the uses that could be made of them, appears in the teachings of Aristotle's student Theophrastus at the Lyceum in ancient Athens in about 350 BC; this is considered the starting point for modern botany. In Europe, this early botanical science was soon overshadowed by a medieval preoccupation with the medicinal properties of plants that lasted more than 1000 years. During this time, the medicinal works of classical antiquity were reproduced in manuscripts and books called herbals. In China and the Arab world, the Greco-Roman work on medicinal plants was preserved and extended.

In Europe the Renaissance of the 14th–17th centuries heralded a scientific revival during which botany gradually emerged from natural history as an independent science, distinct from medicine and agriculture. Herbals were replaced by floras: books that described the native plants of local regions. The invention of the microscope stimulated the study of plant anatomy, and the first carefully designed experiments in plant physiology were performed. With the expansion of trade and exploration beyond Europe, the many new plants being discovered were subjected to an increasingly rigorous process of naming, description, and classification.

Progressively more sophisticated scientific technology has aided the development of contemporary botanical offshoots in the plant sciences, ranging from the applied fields of economic botany (notably agriculture, horticulture and forestry), to the detailed examination of the structure and function of plants and their interaction with the environment over many scales from the large-scale global significance of vegetation and plant communities (biogeography and ecology) through to the small scale of subjects like cell theory, molecular biology and plant biochemistry.

Introduction

Botany (Greek word grass, fodder; Medieval Latin *botanicus* – herb, plant) and zoology are, historically, the core disciplines of biology whose history is closely associated with the natural sciences chemistry, physics and geology. A distinction can be made between botanical science in a pure sense, as the study of plants themselves, and botany as applied science, which studies the human use of plants. Early natural history divided pure botany into three main streams morphology-classification, anatomy and physiology – that is, external form, internal structure, and functional operation. The most obvious topics in applied botany are horticulture, forestry and agriculture although there are many others like weed science, plant pathology, floristry, pharmacognosy, economic botany and ethnobotany which lie outside modern courses in botany. Since the origin of botanical science there has been a progressive increase in the scope of the subject as technology has opened up new techniques and areas of study. Modern molecular systematics, for example, entails the principles and techniques of taxonomy, molecular biology, computer science and more.

Within botany there are a number of sub-disciplines that focus on particular plant groups, each with their own range of related studies (anatomy, morphology etc.). Included here are: phycology (algae), pteridology (ferns), bryology (mosses and liverworts) and palaeobotany (fossil plants) and their histories are treated elsewhere. To this list can be added mycology, the study of fungi, which were once treated as plants, but are now ranked as a unique kingdom.

Ancient Knowledge

Nomadic hunter-gatherer societies passed on, by oral tradition, what they knew (their empirical observations) about the different kinds of plants that they used for food, shelter, poisons, medicines, for ceremonies and rituals etc. The uses of plants by these pre-literate societies influenced the way the plants were named and classified—their uses were embedded in folk-taxonomies, the way they were grouped according to use in everyday communication. The nomadic life-style was drastically changed when settled communities were established in about twelve centres around the world during the Neolithic Revolution which extended from about 10,000 to 2500 years ago depending on the region. With these communities came the development of the technology and skills needed for the domestication of plants and animals and the emergence of the written word provided evidence for the passing of systematic knowledge and culture from one generation to the next.

Plant Lore and Plant Selection

During the Neolithic Revolution plant knowledge increased most obviously through the use of plants for food and medicine. All of today's staple foods were domesticated in prehistoric times as a gradual process of selection of higher-yielding varieties took place, possibly unknowingly, over hundreds to thousands of years. Legumes were cultivated on all continents but cereals made up most of the regular diet: rice in East Asia, wheat and barley in the Middle east, and maize in Central and South America. By Greco-Roman times popular food plants of today, including grapes, apples, figs, and olives, were being listed as named varieties in early manuscripts. Botanical authority William Stearn has observed that *"cultivated plants are mankind's most vital and precious heritage from remote antiquity"*.

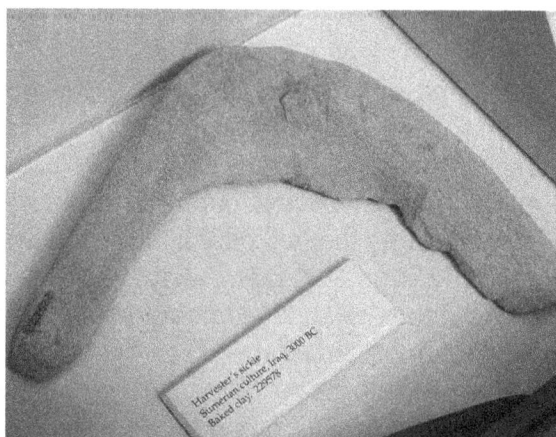

A Sumerian harvester's sickle dated to 3000 BC

It is also from the Neolithic, in about 3000 BC, that we glimpse the first known illustrations of plants and read descriptions of impressive gardens in Egypt. However protobotany, the first pre-scientific written record of plants, did not begin with food; it was born out of the medicinal literature of Egypt, China, Mesopotamia and India. Botanical historian Alan Morton notes that agriculture was the occupation of the poor and uneducated, while medicine was the realm of socially influential shamans, priests, apothecaries, magicians and physicians, who were more likely to record their knowledge for posterity.

Early Botany

Ancient India

An early example of ancient Indian plant classification is found in the Rigveda, a collection of Vedic Sanskrit hymns from about 3700–3100 BP. Plants are divided into *vrska* (trees), *osadhi* (herbs useful to humans) and *virudha* (creepers), with further subdivisions. The sacred Hindu text Atharvaveda divides plants into eight classes: *visakha* (spreading branches), *manjari* (leaves with long clusters), *sthambini* (bushy plants), *prastanavati* (which expands); *ekasrnga* (those with monopodial growth), *pratanavati* (creeping plants), *amsumati* (with many stalks), and *kandini* (plants with knotty joints). The Taittiriya Samhita classifies the plant kingdom into *vrksa, vana* and *druma* (trees), *visakha* (shrubs with spreading branches), *sasa* (herbs), *amsumali* (spreading plant), *vratati* (climber), *stambini* (bushy plant), *pratanavati* (creeper), and *alasala* (spreading on the ground). Other examples of early Indian taxonomy include Manusmriti, the Law book of Hindus, which classifies plants into eight major categories. Elaborate taxonomies also occur in the Charaka Samhitā, Sushruta Samhita and Vaisesika.

Ancient China

In ancient China lists of different plants and herb concoctions for pharmaceutical purposes date back to at least the time of the Warring States (481 BC-221 BC). Many Chinese writers over the centuries contributed to the written knowledge of herbal pharmaceutics. The Han Dynasty (202 BC-220 AD) includes the notable work of the Huangdi Neijing and the famous pharmacologist Zhang Zhongjing. There were also the 11th century scientists and statesmen Su Song and Shen Kuo who compiled learned treatises on natural history, emphasising herbal medicine.

Theophrastus and the Origin of Botanical Science

"School of Athens"
Fresco in Apostolic Palace, Rome, Vatican City, by Raphael 1509-1510

Ancient Athens, of the 6th century BC, was the busy trade centre at the confluence of Egyptian, Mesopotamian and Minoan cultures at the height of Greek colonisation of the Mediterranean. The philosophical thought of this period ranged freely through many subjects. Empedocles (490–430 BC) foreshadowed Darwinian evolutionary theory in a crude formulation of the mutability of species and natural selection. The physician Hippocrates (460–370 BC) avoided the prevailing superstition of his day and approached healing by close observation and the test of experience. At this time a genuine non-anthropocentric curiosity about plants emerged. The major works written about plants extended beyond the description of their medicinal uses to the topics of plant geography, morphology, physiology, nutrition, growth and reproduction.

Foremost among the scholars studying botany was Theophrastus of Eressus who has been frequently referred to as the "Father of Botany". He was a student and close friend of Aristotle (384–322 BC) and succeeded him as head of the Lyceum (an educational establishment like a modern university) in Athens with its tradition of peripatetic philosophy. Aristotle's special treatise on plants is now lost, although there are many botanical observations scattered throughout his other writings (these have been assembled by Christian Wimmer in *Phytologiae Aristotelicae Fragmenta*, 1836) but they give little insight into his botanical thinking. The Lyceum prided itself in a tradition of systematic observation of causal connections, critical experiment and rational theorizing. Theophrastus challenged the superstitious medicine employed by the physicians of his day, called rhizotomi, and also the control over medicine exerted by priestly authority and tradition. Together with Aristotle he had tutored Alexander the Great whose military conquests were carried out with all the scientific resources of the day, the Lyceum garden probably containing many botanical trophies collected during his campaigns as well as other explorations in distant lands. It was in this garden where he gained much of his plant knowledge.

Theophrastus's major botanical works were the *Enquiry into Plants* (*Historia Plantarum*) and *Causes of Plants* (*Causae Plantarum*) which were his lecture notes for the Lyceum. The opening sentence of the *Enquiry* reads like a botanical manifesto: "*We must consider the distinctive characters and the general nature of plants from the point of view of their morphology, their behaviour under external conditions, their mode of generation and the whole course of their life*".

The *Enquiry* is 9 books of "applied" botany dealing with the forms and classification of plants and economic botany, examining the techniques of agriculture (relationship of crops to soil, climate, water and habitat) and horticulture. He described some 500 plants in detail, often including descriptions of habitat and geographic distribution, and he recognised some plant groups that can be recognised as modern-day plant families. Some names he used, like *Crataegus*, *Daucus* and *Asparagus* have persisted until today. His second book *Causes of Plants* covers plant growth and reproduction (akin to modern physiology). Like Aristotle he grouped plants into "trees", "under-shrubs", "shrubs" and "herbs" but he also made several other important botanical distinctions and observations. He noted that plants could be annuals, perennials and biennials, they were also either monocotyledons or dicotyledons and he also noticed the difference between determinate and indeterminate growth and details of floral structure including the degree of fusion of the petals, position of the ovary and more. These lecture notes of Theophrastus comprise the first clear exposition of the rudiments of plant anatomy, physiology, morphology and ecology — presented in a way that would not be matched for another eighteen centuries.

Statue of Theophrastus 371–287 BC "Father of Botany"
Palermo Botanic Gardens

Meanwhile, the study of medicinal plants was not being neglected and a full synthesis of ancient Greek pharmacology was compiled in *Materia Medica* c. 60 AD by Pedanius Dioscorides (c. 40-90 AD) who was a Greek physician with the Roman army. This work proved to be the definitive text on medicinal herbs, both oriental and occidental, for fifteen hundred years until the dawn of the European Renaissance being slavishly copied again and again throughout this period. Though rich in medicinal information with descriptions of about 600 medicinal herbs, the botanical content of the work was extremely limited.

Ancient Rome

The Romans contributed little to the foundations of botanical science laid by the ancient Greeks, but made a sound contribution to our knowledge of applied botany as agriculture. In works titled *De Re Rustica* four Roman writers contributed to a compendium *Scriptores Rei Rusticae*, published from the Renaissance on, which set out the principles and practice of agriculture. These authors were Cato (234–149 BC), Varro (116–27 BC) and, in particular, Columella (4–70 AD) and

Palladius (4th century AD). Roman encyclopaedist Pliny the Elder (23–79 AD) deals with plants in Books 12 to 26 of his 37-volume highly influential work *Naturalis Historia* in which he frequently quotes Theophrastus but with a lack of botanical insight although he does, nevertheless, draw a distinction between true botany on the one hand, and farming and medicine on the other.

It is estimated that at the time of the Roman Empire between 1300 and 1400 plants had been recorded in the West.

Medieval Knowledge

Medicinal Plants of the Early Middle Ages

An Arabic copy of Avicenna's *Canon of Medicine* dated 1593

In Western Europe, after Theophrastus, botany passed through a bleak period of 1800 years when little progress was made and, indeed, many of the early insights were lost. As Europe entered the Middle Ages (5th to 15th centuries), a period of disorganised feudalism and indifference to learning, China, India and the Arab world enjoyed a golden age. Chinese philosophy had followed a similar path to that of the ancient Greeks. The Chinese dictionary-encyclopaedia Erh Ya probably dates from about 300 BC and describes about 334 plants classed as trees or shrubs, each with a common name and illustration. Between 100 and 1700 AD many new works on pharmaceutical botany were produced including encyclopaedic accounts and treatises compiled for the Chinese imperial court. These were free of superstition and myth with carefully researched descriptions and nomenclature; they included cultivation information and notes on economic and medicinal uses — and even elaborate monographs on ornamental plants. But there was no experimental method and no analysis of the plant sexual system, nutrition, or anatomy.

The 400-year period from the 9th to 13th centuries AD was the Islamic Renaissance, a time when Islamic culture and science thrived. Greco-Roman texts were preserved, copied and extended although new texts always emphasised the medicinal aspects of plants. Kurdish biologist Ābu Ḥanīfah Āḥmad ibn Dawūd Dīnawarī (828–896 AD) is known as the founder of Arabic botany; his *Kitâb al-nabât* ('Book of Plants') describes 637 species, discussing plant development from germination to senescence and including details of flowers and fruits. The Mutazilite philosopher and physician

Ibn Sina (Avicenna) (c. 980–1037 AD) was another influential figure, his *The Canon of Medicine* being a landmark in the history of medicine treasured until the Enlightenment.

In India simple artificial plant classification systems of the Rigveda, Atharvaveda and Taittiriya Samhita became more botanical with the work of Parashara (c. 400 – c. 500 AD), the author of *Vṛksayurveda* (the science of life of trees). He made close observations of cells and leaves and divided plants into Dvimatrka (Dicotyledons) and Ekamatrka (Monocotyledons). The dicotyledons were further classified into groupings (ganas) akin to modern floral families: *Samiganiya* (Fabaceae), *Puplikagalniya* (Rutaceae), *Svastikaganiya* (Cruciferae), *Tripuspaganiya* (Cucurbitaceae), *Mallikaganiya* (Apocynaceae), and *Kurcapuspaganiya* (Asteraceae). Important medieval Indian works of plant physiology include the *Prthviniraparyam* of Udayana, *Nyayavindutika* of Dharmottara, *Saddarsana-samuccaya* of Gunaratna, and *Upaskara* of Sankaramisra.

The Silk Road

Following the fall of Constantinople (1453), the newly expanded Ottoman Empire welcomed European embassies in its capital, which in turn became the sources of plants from those regions to the east which traded with the empire. In the following century twenty times as many plants entered Europe along the Silk Road as had been transported in the previous two thousand years, mainly as bulbs. Others were acquired primarily for their alleged medicinal value. Initially Italy benefited from this new knowledge, especially Venice, which traded extensively with the East. From there these new plants rapidly spread to the rest of Western Europe.

The Age of Herbals

Dioscorides', *De Materia Medica*, Byzantium, 15th century

In the European Middle Ages of the 15th and 16th centuries the lives of European citizens were based around agriculture but when printing arrived, with movable type and woodcut illustrations, it was not treatises on agriculture that were published, but lists of medicinal plants with descriptions of their properties or "virtues". These first plant books, known as herbals showed that botany was still a part of medicine, as it had been for most of ancient history. Authors of herbals were often curators of university gardens, and most herbals were derivative compilations of classic texts, especially *De Materia Medica*. However, the need for accurate and detailed plant descriptions meant that some herbals were more botanical than medicinal. German Otto Brunfels's (1464–1534) *Herbarum Vivae Icones* (1530) contained descriptions of about 47 species new to science combined with accurate illustrations. His fellow countryman Hieronymus Bock's (1498–1554) *Kreutterbuch* of 1539 described plants he found in nearby woods and fields and these were illustrated in the 1546 edition. However, it was Valerius Cordus (1515–1544) who pioneered the formal botanical description that detailed both flowers and fruits, some anatomy including the number of chambers in the ovary, and the type of ovule placentation. He also made observations on pollen and distinguished between inflorescence types. His five-volume *Historia Plantarum* was published about 18 years after his early death aged 29 in 1561-1563. In Holland Rembert Dodoens (1517–1585), in *Stirpium Historiae* (1583), included descriptions of many new species from the Netherlands in a scientific arrangement and in England William Turner (1515–1568) in his *Libellus De Re Herbaria Novus* (1538) published names, descriptions and localities of many native British plants.

Herbals contributed to botany by setting in train the science of plant description, classification, and botanical illustration. Up to the 17th century botany and medicine were one and the same but those books emphasising medicinal aspects eventually omitted the plant lore to become modern pharmacopoeias; those that omitted the medicine became more botanical and evolved into the modern compilations of plant descriptions we call Floras. These were often backed by specimens deposited in a herbarium which was a collection of dried plants that verified the plant descriptions given in the Floras. The transition from herbal to Flora marked the final separation of botany from medicine.

The Renaissance and Age of Enlightenment (1550–1800)

The revival of learning during the European Renaissance renewed interest in plants. The church, feudal aristocracy and an increasingly influential merchant class that supported science and the arts, now jostled in a world of increasing trade. Sea voyages of exploration returned botanical treasures to the large public, private, and newly established botanic gardens, and introduced an eager population to novel crops, drugs and spices from Asia, the East Indies and the New World.

The number of scientific publications increased. In England, for example, scientific communication and causes were facilitated by learned societies like Royal Society (founded in 1660) and the Linnaean Society (founded in 1788): there was also the support and activities of botanical institutions like the Jardin du Roi in Paris, Chelsea Physic Garden, Royal Botanic Gardens Kew, and the Oxford and Cambridge Botanic Gardens, as well as the influence of renowned private gardens and wealthy entrepreneurial nurserymen. By the early 17th century the number of plants described in Europe had risen to about 6000. The 18th century Enlightenment values of reason and science coupled with new voyages to distant lands instigating another phase of encyclopaedic plant identification, nomenclature, description and illustration, "flower painting" possibly at its best in this period of history. Plant trophies from distant lands decorated the gardens of Europe's powerful and wealthy in a period of enthusiasm for natural history, especially botany (a preoccupation sometimes referred to as "botanophilia") that is never likely to recur.

During the 18th century botany was one of the few sciences considered appropriate for genteel educated women. Around 1760, with the popularization of the Linnaean system, botany became much more widespread among educated women who painted plants, attended classes on plant classification, and collected herbarium specimens although emphasis was on the healing properties of plants rather than plant reproduction which had overtones of sexuality. Women began publishing on botanical topics and children's books on botany appeared by authors like Charlotte

Turner Smith. Cultural authorities argued that education through botany created culturally and scientifically aware citizens, part of the thrust for 'improvement' that characterised the Enlightenment. However, in the early 19th century with the recognition of botany as an official science women were again excluded from the discipline.

Botanical Gardens and Herbaria

A 16th century print of the Botanical Garden of Padova (*Garden of the Simples*)
— the oldest academic botanic garden that is still in its original location

Public and private gardens have always been strongly associated with the historical unfolding of botanical science. Early botanical gardens were physic gardens, repositories for the medicinal plants described in the herbals. As they were generally associated with universities or other academic institutions the plants were also used for study. The directors of these gardens were eminent physicians with an educational role as "scientific gardeners" and it was staff of these institutions that produced many of the published herbals.

The botanical gardens of the modern tradition were established in northern Italy, the first being at Pisa (1544), founded by Luca Ghini (1490–1556). Although part of a medical faculty, the first chair of *materia medica*, essentially a chair in botany, was established in Padua in 1533. Then in 1534, Ghini became Reader in *materia medica* at Bologna University, where Aldrovandi established a similar garden in 1568. Collections of pressed and dried specimens were called a *hortus siccus* (garden of dry plants) and the first accumulation of plants in this way (including the use of a plant press) is attributed to Ghini. Buildings called herbaria housed these specimens mounted on card with descriptive labels. Stored in cupboards in systematic order they could be preserved in perpetuity and easily transferred or exchanged with other institutions, a taxonomic procedure that is still used today.

By the 18th century the physic gardens had been transformed into "order beds" that demonstrated the classification systems that were being devised by botanists of the day — but they also had to accommodate the influx of curious, beautiful and new plants pouring in from voyages of exploration that were associated with European colonial expansion.

From Herbal to Flora

Plant classification systems of the 17th and 18th centuries now related plants to one another and not to man, marking a return to the non-anthropocentric botanical science promoted by Theophrastus over 1500 years before. In England, various herbals in either Latin or English were mainly compilations and translations of continental European works, of limited relevance to the British Isles. This included the rather unreliable work of Gerard (1597). The first systematic attempt to collect information on British plants was that of Thomas Johnson (1629), who was later to issue his own revision of Gerard's work (1633–1636).

However Johnson was not the first apothecary or physician to organise botanical expeditions to systematise their local flora. In Italy Ulysse Aldrovandi (1522 – 1605) organised an expedition to the Sibylline mountains in Umbria in 1557, and compiled a local Flora. He then began to disseminate his findings amongst other European scholars, forming an early network of knowledge sharing "*molti amici in molti luoghi*" (many friends in many places), including Charles de l'Écluse (Clusius) (1526 – 1609) at Montpellier and Jean de Brancion at Malines. Between them they started developing Latin names for plants, in addition to their common names. The exchange of information and specimens between scholars was often associated with the founding of botanical gardens (above), and to this end Aldrovandi founded one of the earliest at his university in Bologna, the Orto Botanico di Bologna in 1568.

In France, Clusius journeyed throughout most of Western Europe, making discoveries in the vegetable kingdom along the way. He compiled Flora of Spain (1576), and Austria and Hungary (1583). He was the first to propose dividing plants into classes. Meanwhile, in Switzerland, from 1554, Conrad Gessner (1516 – 1565) made regular explorations of the Swiss Alps from his native Zurich and discovered many new plants. He proposed that there were groups or genera of plants. He said that each genus was composed of many species and that these were defined by similar flowers and fruits. This principle of organization laid the groundwork for future botanists. He wrote his important *Historia Plantarum* shortly before his death. At Malines, in Flanders he established and maintained the botanical gardens of Jean de Brancion from 1568 to 1573, and first encountered tulips.

This approach coupled with the new Linnaean system of binomial nomenclature resulted in plant encyclopaedias without medicinal information called *Floras* that meticulously described and illustrated the plants growing in particular regions. The 17th century also marked the beginning of experimental botany and application of a rigorous scientific method, while improvements in the microscope launched the new discipline of plant anatomy whose foundations, laid by the careful observations of Englishman Nehemiah Grew and Italian Marcello Malpighi, would last for 150 years.

Botanical Exploration

More new lands were opening up to European colonial powers, the botanical riches being returned to European botanists for description. This was a romantic era of botanical explorers, intrepid plant hunters and gardener-botanists. Significant botanical collections came from: the West Indies (Hans Sloane (1660–1753)); China (James Cunningham); the spice islands of the East Indies (Moluccas, George Rumphius (1627–1702)); China and Mozambique (João de Loureiro (1717–1791)); West Africa (Michel Adanson (1727–1806)) who devised his own classification scheme and forwarded a crude theory of the mutability of species; Canada, Hebrides, Iceland, New Zealand by Captain James Cook's chief botanist Joseph Banks (1743–1820).

Classification and Morphology

Portrait of Carl Linnaeus by Alexander Roslin, 1775

By the middle of the 18th century the botanical booty resulting from the era of exploration was accumulating in gardens and herbaria – and it needed to be systematically catalogued. This was the task of the taxonomists, the plant classifiers.

Plant classifications have changed over time from "artificial" systems based on general habit and form, to pre-evolutionary "natural" systems expressing similarity using one to many characters, leading to post-evolutionary "natural" systems that use characters to infer evolutionary relationships.

Italian physician Andrea Caesalpino (1519–1603) studied medicine and taught botany at the University of Pisa for about 40 years eventually becoming Director of the Botanic Garden of Pisa from 1554 to 1558. His sixteen-volume *De Plantis* (1583) described 1500 plants and his herbarium of 260 pages and 768 mounted specimens still remains. Caesalpino proposed classes based largely on the detailed structure of the flowers and fruit; he also applied the concept of the genus. He was the first to try and derive principles of natural classification reflecting the overall similarities between plants and he produced a classification scheme well in advance of its day. Gaspard Bauhin (1560–1624) produced two influential publications *Prodromus Theatrici Botanici* (1620) and *Pinax* (1623). These brought order to the 6000 species now described and in the latter he used binomials and synonyms that may well have influenced Linnaeus's thinking. He also insisted that taxonomy should be based on natural affinities.

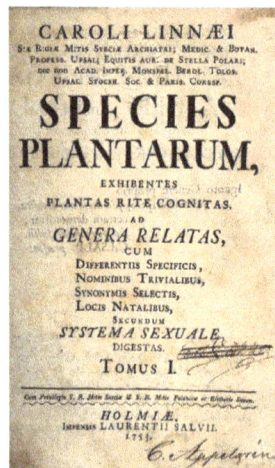

Cover page of *Species Plantarum* of Carl Linnaeus published in 1753

To sharpen the precision of description and classification Joachim Jung (1587–1657) compiled a much-needed botanical terminology which has stood the test of time. English botanist John Ray (1623–1705) built on Jung's work to establish the most elaborate and insightful classification system of the day. His observations started with the local plants of Cambridge where he lived, with the *Catalogus Stirpium circa Cantabrigiam Nascentium* (1860) which later expanded to his *Synopsis Methodica Stirpium Britannicarum*, essentially the first British Flora. Although his *Historia Plantarum* (1682, 1688, 1704) provided a step towards a world Flora as he included more and more plants from his travels, first on the continent and then beyond. He extended Caesalpino's natural system with a more precise definition of the higher classification levels, deriving many modern families in the process, and asserted that all parts of plants were important in classification. He recognised that variation arises from both internal (genotypic) and external environmental (phenotypic) causes and that only the former was of taxonomic significance. He was also among the first experimental physiologists. The *Historia Plantarum* can be regarded as the first botanical synthesis and textbook for modern botany. According to botanical historian Alan Morton, Ray "influenced both the theory and the practice of botany more decisively than any other single person in the latter half of the seventeenth century". Ray's family system was later extended by Pierre Magnol (1638–1715) and Joseph de Tournefort (1656–1708), a student of Magnol, achieved notoriety for his botanical expeditions, his emphasis on floral characters in classification, and for reviving the idea of the genus as the basic unit of classification.

Above all it was Swedish Carl Linnaeus (1707–1778) who eased the task of plant cataloguing. He adopted a sexual system of classification using stamens and pistils as important characters. Among his most important publications were Systema Naturae (1735), Genera Plantarum (1737), and Philosophia Botanica (1751) but it was in his Species Plantarum (1753) that he gave every species a binomial thus setting the path for the future accepted method of designating the names of all organisms. Linnaean thought and books dominated the world of taxonomy for nearly a century. His sexual system was later elaborated by Bernard de Jussieu (1699–1777) whose nephew Antoine-Laurent de Jussieu (1748–1836) extended it yet again to include about 100 orders (present-day families). Frenchman Michel Adanson (1727–1806) in his *Familles des Plantes* (1763, 1764), apart from extending the current system of family names, emphasized that a natural classification must be based on a consideration of all characters, even though these may later be given different emphasis according to their diagnostic value for the particular plant group. Adanson's method has, in essence, been followed to this day.

18th century plant taxonomy bequeathed to the 19th century a precise binomial nomenclature and botanical terminology, a system of classification based on natural affinities, and a clear idea of the ranks of family, genus and species — although the taxa to be placed within these ranks remains, as always, the subject of taxonomic research.

Anatomy

In the first half of the 18th century botany was beginning to move beyond descriptive science into experimental science. Although the microscope was invented in 1590 it was only in the late 17th century that lens grinding by Antony van Leeuwenhoek provided the resolution needed to make major discoveries. Important general biological observations were made by Robert Hooke (1635–1703) but the foundations of plant anatomy were laid by Italian Marcello Malpighi (1628–1694) of the University of Bologna in his *Anatome Plantarum* (1675) and Royal Society Englishman Nehemiah Grew (1628–1711) in his *The Anatomy of Plants Begun* (1671) and *Anatomy of Plants* (1682). These botanists explored what is now called developmental anatomy and morphology by carefully observing, describing and drawing the developmental transition from seed to mature plant, recording stem and wood formation. This work included the discovery and naming of parenchyma and stomata.

Robert Hooke's microscope which he described in the 1665 *Micrographia*:
he coined the biological use of the term *cell*

Physiology

In plant physiology research interest was focused on the movement of sap and the absorption of substances through the roots. Jan Helmont (1577–1644) by experimental observation and calculation, noted that the increase in weight of a growing plant cannot be derived purely from the soil, and concluded it must relate to water uptake. Englishman Stephen Hales (1677–1761) established by quantitative experiment that there is uptake of water by plants and a loss of water by transpiration and that this is influenced by environmental conditions: he distinguished "root pressure", "leaf suction" and "imbibition" and also noted that the major direction of sap flow in woody tissue is upward. His results were published in *Vegetable Staticks* (1727) He also noted that "air makes a very considerable part of the substance of vegetables". English chemist Joseph Priestley (1733–1804) is noted for his discovery of oxygen (as now called) and its production by plants. Later Jan Ingenhousz (1730–1799) observed that only in sunlight do the green parts of plants absorb air and release oxygen, this being more rapid in bright sunlight while, at night, the air (CO_2) is released from all parts. His results were published in *Experiments upon vegetables* (1779) and with this the foundations for 20th century studies of carbon fixation were laid. From his observations he sketched the cycle of carbon in nature even though the composition of carbon dioxide was yet to be resolved. Studies in plant nutrition had also progressed. In 1804 Nicolas-Théodore de Saussure's (1767–1845) *Recherches Chimiques sur la Végétation* was an exemplary study of scientific exactitude that demonstrated the similarity of respiration in both plants and animals, that the fixation of carbon dioxide includes water, and that just minute amounts of salts and nutrients (which he analysed in chemical detail from plant ash) have a powerful influence on plant growth.

Plant Sexuality

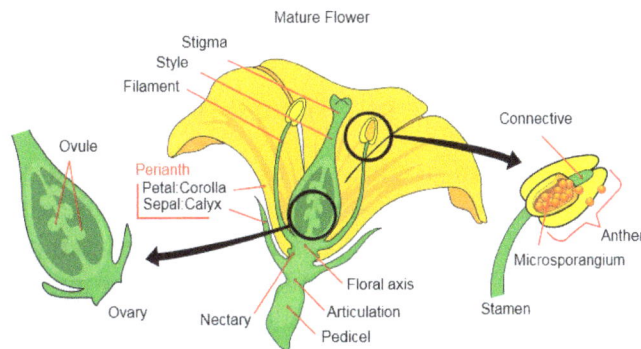

Diagram showing the sexual parts of a mature flower

It was Rudolf Camerarius (1665–1721) who was the first to establish plant sexuality conclusively by experiment. He declared in a letter to a colleague dated 1694 and titled *De Sexu Plantarum Epistola* that "no ovules of plants could ever develop into seeds from the female style and ovary without first being prepared by the pollen from the stamens, the male sexual organs of the plant".

Much was learned about plant sexuality by unravelling the reproductive mechanisms of mosses, liverworts and algae. In his *Vergleichende Untersuchungen* of 1851 Wilhelm Hofmeister (1824–1877) starting with the ferns and bryophytes demonstrated that the process of sexual reproduction in plants entails an "alternation of generations" between sporophytes and gametophytes. This initiated the new field of comparative morphology which, largely through the combined work of William Farlow (1844–1919), Nathanael Pringsheim (1823–1894), Frederick Bower, Eduard Strasburger and others, established that an "alternation of generations" occurs throughout the plant kingdom.

Some time later the German academic and natural historian Joseph Kölreuter (1733–1806) extended this work by noting the function of nectar in attracting pollinators and the role of wind and insects in pollination. He also produced deliberate hybrids, observed the microscopic structure of pollen grains and how the transfer of matter from the pollen to the ovary inducing the formation of the embryo.

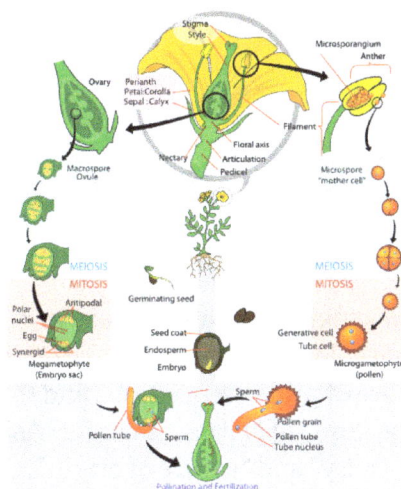

Angiosperm (flowering plant) life cycle showing alternation of generations

One hundred years after Camerarius, in 1793, Christian Sprengel (1750–1816) broadened the understanding of flowers by describing the role of nectar guides in pollination, the adaptive floral mechanisms used for pollination, and the prevalence of cross-pollination, even though male and female parts are usually together on the same flower.

Nineteenth-century Foundations of Modern Botany

In about the mid-19th century scientific communication changed. Until this time ideas were largely exchanged by reading the works of authoritative individuals who dominated in their field: these were often wealthy and influential "gentlemen scientists". Now research was reported by the publication of "papers" that emanated from research "schools" that promoted the questioning of conventional wisdom. This process had started in the late 18th century when specialist journals began to appear. Even so, botany was greatly stimulated by the appearance of the first "modern" textbook, Matthias Schleiden's (1804–1881) *Grundzüge der Wissenschaftlichen Botanik*, published in English in 1849 as *Principles of Scientific Botany*. By 1850 an invigorated organic chemistry had revealed the structure of many plant constituents. Although the great era of plant classification had now passed the work of description continued. Augustin de Candolle (1778–1841) succeeded Antoine-Laurent de Jussieu in managing the botanical project *Prodromus Systematis Naturalis Regni Vegetabilis* (1824–1841) which involved 35 authors: it contained all the dicotyledons known in his day, some 58000 species in 161 families, and he doubled the number of recognized plant families, the work being completed by his son Alphonse (1806–1893) in the years from 1841 to 1873.

Plant Geography and Ecology

Alexander von Humboldt 1769–1859 painted by Joseph Stieler in 1843

The opening of the 19th century was marked by an increase in interest in the connection between climate and plant distribution. Carl Willdenow (1765–1812) examined the connection between seed dispersal and distribution, the nature of plant associations and the impact of geological history. He noticed the similarities between the floras of N America and N Asia, the Cape and Australia, and he explored the ideas of "centre of diversity" and "centre of origin". German Alexander von Humboldt (1769–1859) and Frenchman Aime Bonpland (1773–1858) published a massive and highly influential 30 volume work on their travels; Robert Brown (1773–1852) noted the similarities between the floras of S Africa, Australia and India, while Joakim Schouw (1789–1852) explored more deeply than anyone else the influence on plant distribution of temperature, soil factors, especially soil water, and light, work that was continued by Alphonse de Candolle (1806–1893). Joseph Hooker (1817–1911) pushed the boundaries of floristic studies with his work on Antarctica, India and the Middle East with special attention to endemism. August Grisebach (1814–1879) in *Die Vegetation der Erde* (1872) examined physiognomy in relation to climate and in America geographic studies were pioneered by Asa Gray (1810–1888).

Physiological plant geography, perhaps more familiarly termed ecology, emerged from floristic biogeography in the late 19th century as environmental influences on plants received greater recognition. Early work in this area was synthesised by Danish professor Eugenius Warming (1841–1924) in his book *Plantesamfund* (Ecology of Plants, generally taken to mark the beginning of modern ecology) including new ideas on plant communities, their adaptations and environmental influences. This was followed by another grand synthesis, the *Pflanzengeographie auf Physiologischer Grundlage* of Andreas Schimper (1856–1901) in 1898 (published in English in 1903 as Plant-geography upon a physiological basis translated by W. R. Fischer, Oxford: Clarendon press, 839 pp.)

Anatomy

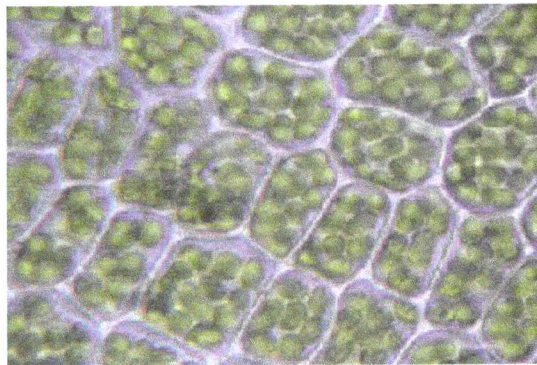

Plant cells with visible chloroplasts

During the 19th century German scientists led the way towards a unitary theory of the structure and life-cycle of plants. Following improvements in the microscope at the end of the 18th century, Charles Mirbel (1776–1854) in 1802 published his *Traité d'Anatomie et de Physiologie Végétale* and Johann Moldenhawer (1766–1827) published *Beyträge zur Anatomie der Pflanzen* (1812) in which he describes techniques for separating cells from the middle lamella. He identified vascular and parenchymatous tissues, described vascular bundles, observed the cells in the cambium, and interpreted tree rings. He found that stomata were composed of pairs of cells, rather than a single cell with a hole.

Anatomical studies on the stele were consolidated by Carl Sanio (1832–1891) who described the secondary tissues and meristem including cambium and its action. Hugo von Mohl (1805–1872) summarized work in anatomy leading up to 1850 in *Die Vegetabilische Zelle* (1851) but this work was later eclipsed by the encyclopaedic comparative anatomy of Heinrich Anton de Bary in 1877. An overview of knowledge of the stele in root and stem was completed by Van Tieghem (1839–

1914) and of the meristem by Karl Nägeli (1817–1891). Studies had also begun on the origins of the carpel and flower that continue to the present day.

Water Relations

The riddle of water and nutrient transport through the plant remained. Physiologist Von Mohl explored solute transport and the theory of water uptake by the roots using the concepts of cohesion, transpirational pull, capillarity and root pressure. German dominance in the field of physiology was underlined by the publication of the definitive textbook on plant physiology synthesising the work of this period, Sach's *Vorlesungen über Pflanzenphysiologie* of 1882. There were, however, some advances elsewhere such as the early exploration of geotropism (the effect of gravity on growth) by Englishman Thomas Knight, and the discovery and naming of osmosis by Frenchman Henri Dutrochet (1776–1847).

Cytology

The cell nucleus was discovered by Robert Brown in 1831. Demonstration of the cellular composition of all organisms, with each cell possessing all the characteristics of life, is attributed to the combined efforts of botanist Matthias Schleiden and zoologist Theodor Schwann (1810–1882) in the early 19th century although Moldenhawer had already shown that plants were wholly cellular with each cell having its own wall and Julius von Sachs had shown the continuity protoplasm between cell walls.

From 1870 to 1880 it became clear that cell nuclei are never formed anew but always derived from the substance of another nucleus. In 1882 Flemming observed the longitudinal splitting of chromosomes in the dividing nucleus and concluded that each daughter nucleus received half of each of the chromosomes of the mother nucleus: then by the early 20th century it was found that the number of chromosomes in a given species is constant. With genetic continuity confirmed and the finding by Eduard Strasburger that the nuclei of reproductive cells (in pollen and embryo) have a reducing division (halving of chromosomes, now known as meiosis) the field of heredity was opened up. By 1926 Thomas Morgan was able to outline a theory of the gene and its structure and function. The form and function of plastids received similar attention, the association with starch being noted at an early date. With observation of the cellular structure of all organisms and the process of cell division and continuity of genetic material, the analysis of the structure of protoplasm and the cell wall as well as that of plastids and vacuoles – what is now known as cytology, or cell theory became firmly established.

Later, the cytological basis of the gene-chromosome theory of heredity extended from about 1900–1944 and was initiated by the rediscovery of Gregor Mendel's (1822–1884) laws of plant heredity first published in 1866 in *Experiments on Plant Hybridization* and based on cultivated pea, *Pisum sativum*: this heralded the opening up of plant genetics. The cytological basis for gene-chromosome theory was explored through the role of polyploidy and hybridization in speciation and it was becoming better understood that interbreeding populations were the unit of adaptive change in biology.

Developmental Morphology and Evolution

Until the 1860s it was believed that species had remained unchanged through time: each biological form was the result of an independent act of creation and therefore absolutely distinct and immutable. But the hard reality of geological formations and strange fossils needed scientific explanation. Charles Darwin's *Origin of Species* (1859) replaced the assumption of constancy with the theory of descent with modification. Phylogeny became a new principle as "natural" classifications became classifications reflecting, not just similarities, but evolutionary relationships. Wilhelm Hofmeister established that there was a similar pattern of organization in all plants expressed through the alternation of generations and extensive homology of structures.

Polymath German intellect Johann Goethe (1749–1832) had interests and influence that extended into botany. In *Die Metamorphose der Pflanzen* (1790) he provided a theory of plant morphology (he coined the word "morphology") and he included within his concept of "metamorphosis" modification during evolution, thus linking comparative morphology with phylogeny. Though the botanical basis of his work has been challenged there is no doubt that he prompted discussion and research on the origin and function of floral parts. His theory probably stimulated the opposing views of German botanists Alexander Braun (1805–1877) and Matthias Schleiden who applied the experimental method to the principles of growth and form that were later extended by Augustin de Candolle (1778–1841).

Carbon Fixation (Photosynthesis)

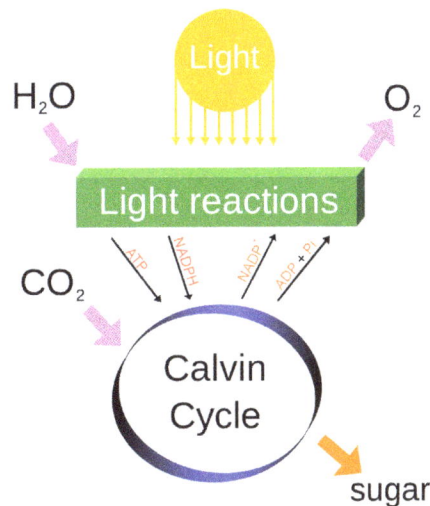

Photosynthesis splits water to liberate O_2 and fixes CO_2 into sugar

At the start of the 19th century the idea that plants could synthesise almost all their tissues from atmospheric gases had not yet emerged. The energy component of photosynthesis, the capture and storage of the Sun's radiant energy in carbon bonds (a process on which all life depends) was first elucidated in 1847 by Mayer, but the details of how this was done would take many more years. Chlorophyll was named in 1818 and its chemistry gradually determined, to be finally resolved in the early 20th century. The mechanism of photosynthesis remained a mystery until the mid-19th century when Sachs, in 1862, noted that starch was formed in green cells only in the presence of light and in 1882 he confirmed carbohydrates as the starting point for all other organic compounds in plants. The connection between the pigment chlorophyll and starch production was finally made in 1864 but tracing the precise biochemical pathway of starch formation did not begin until about 1915.

Nitrogen Fixation

Significant discoveries relating to nitrogen assimilation and metabolism, including ammonification, nitrification and nitrogen fixation (the uptake of atmospheric nitrogen by symbiotic soil microorganisms) had to wait for advances in chemistry and bacteriology in the late 19th century and this was followed in the early 20th century by the elucidation of protein and amino-acid synthesis and their role in plant metabolism. With this knowledge it was then possible to outline the global nitrogen cycle.

Twentieth Century

Thin layer chromatography is used to separate components of chlorophyll

20th century science grew out of the solid foundations laid by the breadth of vision and detailed experimental observations of the 19th century. A vastly increased research force was now rapidly extending the horizons of botanical knowledge at all levels of plant organization from molecules to global plant ecology. There was now an awareness of the unity of biological structure and function at the cellular and biochemical levels of organisation. Botanical advance was closely associated with advances in physics and chemistry with the greatest advances in the 20th century mainly relating to the penetration of molecular organization. However, at the level of plant communities it would take until mid century to consolidate work on ecology and population genetics. By 1910 experiments using labelled isotopes were being used to elucidate plant biochemical pathways, to open the line of research leading to gene technology. On a more practical level research funding was now becoming available from agriculture and industry.

Molecules

In 1903 Chlorophylls a and b were separated by thin layer chromatography then, through the 1920s and 1930s, biochemists, notably Hans Krebs (1900–1981) and Carl (1896–1984) and Gerty Cori (1896–1957) began tracing out the central metabolic pathways of life. Between the 1930s and 1950s it was determined that ATP, located in mitochondria, was the source of cellular chemical energy and the constituent reactions of photosynthesis were progressively revealed. Then, in 1944 DNA was extracted for the first time. Along with these revelations there was the discovery of plant hormones or "growth substances", notably auxins, (1934) gibberellins (1934) and cytokinins (1964) and the effects of photoperiodism, the control of plant processes, especially flowering, by the relative lengths of day and night.

Following the establishment of Mendel's laws, the gene-chromosome theory of heredity was confirmed by the work of August Weismann who identified chromosomes as the hereditary material. Also, in observing the halving of the chromosome number in germ cells he anticipated work to follow on the details of meiosis, the complex process of redistribution of hereditary material that occurs in the germ cells. In the 1920s and 1930s population genetics combined the theory of evolution with Mendelian genetics to produce the modern synthesis. By the mid-1960s the molecular basis of metabolism and reproduction was firmly established through the new discipline of molecular biology. Genetic engineering, the insertion of genes into a host cell for cloning, began in the 1970s with the invention of recombinant DNA techniques and its commercial applications applied to agricultural crops followed in the 1990s. There was now the potential to identify organisms by molecular "fingerprinting" and to estimate the times in the past when critical evolutionary changes had occurred through the use of "molecular clocks".

Computers, Electron Microscopes and Evolution

Electron microscope constructed by Ernst Ruska in 1933

Increased experimental precision combined with vastly improved scientific instrumentation was opening up exciting new fields. In 1936 Alexander Oparin (1894–1980) demonstrated a possible mechanism for the synthesis of organic matter from inorganic molecules. In the 1960s it was determined that the Earth's earliest life-forms treated as plants, the cyanobacteria known as stromatolites, dated back some 3.5 billion years.

Mid-century transmission and scanning electron microscopy presented another level of resolution to the structure of matter, taking anatomy into the new world of "ultrastructure".

New and revised "phylogenetic" classification systems of the plant kingdom were produced, perhaps the most notable being that of August Eichler (1839–1887), and the massive 23 volume *Die natürlichen Pflanzenfamilien* of Adolf Engler (1844–1930) & Karl Prantl (1849–1893) published over the period 1887 and 1915. Taxonomy based on gross morphology was now being supplemented by using characters revealed by pollen morphology, embryology, anatomy, cytology, serology, macromolecules and more. The introduction of computers facilitated the rapid analysis of large data sets used for numerical taxonomy (also called taximetrics or phenetics). The emphasis on truly natural phylogenies spawned the disciplines of cladistics and phylogenetic systematics. The grand taxonomic synthesis *An Integrated System of Classification of Flowering Plants* (1981) of American Arthur Cronquist (1919–1992) was superseded when, in 1998, the Angiosperm Phylogeny Group published a phylogeny of flowering plants based on the analysis of DNA sequences using the techniques of the new molecular systematics which was resolving questions concerning the earliest evolutionary branches of the angiosperms (flowering plants). The exact relationship of fungi to plants had for some time been uncertain. Several lines of evidence pointed to fungi being different from plants, animals and bacteria – indeed, more closely related to animals than plants. In the 1980s-90s molecular analysis revealed an evolutionary divergence of fungi from other organisms about 1 billion years ago – sufficient reason to erect a unique kingdom separate from plants.

Biogeography and Ecology

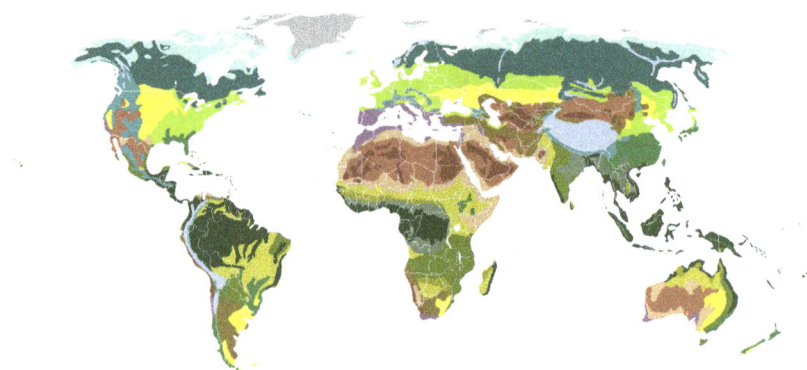

Map of terrestrial biomes classified by vegetation type

The publication of Alfred Wegener's (1880–1930) theory of continental drift 1912 gave additional impetus to comparative physiology and the study of biogeography while ecology in the 1930s contributed the important ideas of plant community, succession, community change, and energy flows. From 1940 to 1950 ecology matured to become an independent discipline as Eugene Odum (1913–2002) formulated many of the concepts of ecosystem ecology, emphasising relationships between groups of organisms (especially material and energy relationships) as key factors in the field. Building on the extensive earlier work of Alphonse de Candolle, Nikolai Vavilov (1887–1943) from 1914 to 1940 produced accounts of the geography, centres of origin, and evolutionary history of economic plants.

Twenty-first Century

In reviewing the sweep of botanical history it is evident that, through the power of the scientific method, most of the basic questions concerning the structure and function of plants have, in principle, been resolved. Now the distinction between pure and applied botany becomes blurred as our historically accumulated botanical wisdom at all levels of plant organisation is needed (but especially at the molecular and global levels) to improve human custodianship of planet earth. The most urgent unanswered botanical questions now relate to the role of plants as primary producers in the global cycling of life's basic ingredients: energy, carbon, hydrogen, oxygen, and nitrogen, and ways that our plant stewardship can help address the global environmental issues of resource management, conservation, human food security, biologically invasive organisms, carbon sequestration, climate change, and sustainability.

History of Plant Systematics

The *Vienna Dioscurides* manuscript of *De Materia Medica*, from the early sixth century, is one of the oldest herbals in existence. Dioscorides wrote the book between 50 and 70 AD

The history of plant systematics—the biological classification of plants—stretches from the work of ancient Greek to modern evolutionary biologists. As a field of science, plant systematics came into being only slowly, early plant lore usually being treated as part of the study of medicine. Later, classification and description was driven by natural history and natural theology. Until the advent of the theory of evolution, nearly all classification was based on the scala naturae. The professionalization of botany in the 18th and 19th century marked a shift toward more holistic classification methods, eventually based on evolutionary relationships.

Antiquity

The peripatetic philosopher Theophrastus (372–287 BC), as a student of Aristotle in Ancient Greece, wrote *Historia Plantarum*, the earliest surviving treatise on plants, where he listed the names of over 500 plant species. He did not articulate a formal classification scheme, but relied on the common groupings of folk taxonomy combined with growth form: tree shrub; undershrub; or herb.

The *De Materia Medica* of Dioscorides was an important early compendium of plant descriptions (over five hundred), classifying plants chiefly by their medicinal effects; it was in use from its publication in the 1st century until the 16th century, making it the major herbal throughout the Middle Ages.

Early Modern Period

In the 16th century, works by Otto Brunfels, Hieronymus Bock, and Leonhart Fuchs helped to revive interest in natural history based on first-hand observation; Bock in particular included environmental and life cycle information in his descriptions. With the influx of exotic species in the Age of Exploration, the number of known species expanded rapidly, but most authors were far more interested in the medicinal properties of individual plants than an overarching classification system. Later influential Renaissance books include those of Caspar Bauhin and Andrea Cesalpino. Bauhin described over 6000 plants, which he arranged into 12 books and 72 sections based on a wide range of common characteristics. Cesalpino based his system on the structure of the organs of fructification, using the Aristotelian technique of logical division.

In the late 17th century, the most influential classification schemes were those of English botanist and natural theologian John Ray and French botanist Joseph Pitton de Tournefort. Ray, who listed over 18,000 plant species in his works, is credited with establishing the monocot/dicot division and some of his groups — mustards, mints, legumes and grasses — stand today (though under modern family names). Tournefort used an artificial system based on logical division which was widely adopted in France and elsewhere in Europe up until Linnaeus.

The book that had an enormous accelerating effect on the science of plant systematics was *Species Plantarum* (1753) by Linnaeus. It presented a complete list of the plant species then known to Europe, ordered for the purpose of easy identification using the number and arrangement of the male and female sexual organs of the plants. Of the groups in this book, the highest rank that continues to be used today is the genus. The consistent use of binomial nomenclature along with a complete listing of all plants provided a huge stimulus for the field.

Although meticulous, the classification of Linnaeus served merely as an identification manual; it was based on phenetics and did not regard evolutionary relationships among species. It assumed that plant species were given by God and that what remained for humans was to recognise them and use them (A Christian reformulation of the *scala naturae* or *Great Chain of Being*). Linnaeus was quite aware that the arrangement of species in the *Species Plantarum* was not a natural system, i.e. did not express relationships. However he did present some ideas of plant relationships elsewhere.

Modern and Contemporary Periods

Significant contributions to plant classification came from de Jussieu (inspired by the work of Adanson) in 1789 and the early nineteenth century saw the start of work by de Candolle, culminating in the *Prodromus*.

A major influence on plant systematics was the theory of evolution (Charles Darwin published *Origin of Species* in 1859), resulting in the aim to group plants by their phylogenetic relationships. To this was added the interest in plant anatomy, aided by the use of the light microscope and the rise of chemistry, allowing the analysis of secondary metabolites.

Currently, the strict use of epithets in botany, although regulated by international codes, is considered unpractical and outdated. The very notion of species, the fundamental classification unit, is often up to subjective intuition and thus can not be well defined. As a result, estimate of the total number of existing "species" (ranging from 2 million to 100 million) becomes a matter of preference.

While scientists have agreed for some time that a functional and objective classification system must reflect actual evolutionary processes and genetic relationships, the technological means for creating such a system did not exist until recently. In the 1990s DNA technology saw immense progress, resulting in unprecedented accumulation of DNA sequence data from various genes present in compartments of plant cells. In 1998 a ground-breaking classification of the angiosperms (the APG system) consolidated molecular phylogenetics (and especially cladistics or phylogenetic systematics) as the best available method. For the first time relatedness could be measured in real terms, namely similarity of the molecules comprising the genetic code.

References

- McCutcheon, A. R.; Ellis, S. M.; Hancock, R. E.; Towers, G. H. (1992-10-01). "Antibiotic screening of medicinal plants of the British Columbian native peoples". Journal of Ethnopharmacology. 37 (3): 213–223. ISSN 0378-8741. PMID 1453710. doi:10.1016/0378-8741(92)90036-q

- "Research confirms Native American use of sweetgrass as bug repellent". Washington Post. Retrieved 2016-05-05

- Sutton, David; Robert Huxley (editor) (2007). "Pedanios Dioscorides: Recording the Medicinal Uses of Plants". The Great Naturalists. London: Thames & Hudson, with the Natural History Museum. pp. 32–37. ISBN 978-0-500-25139-3

- TeachEthnobotany (2012-06-12), Cultivation of peyote by Native Americans: Past, present and future, retrieved 2016-05-05

- Delcourt, Paul A.; Delcourt, Hazel R.; Cridlebaugh, Patricia A.; Chapman, Jefferson (1986-05-01). "Holocene ethnobotanical and paleoecological record of human impact on vegetation in the Little Tennessee River Valley, Tennessee". Quaternary Research. 25 (3): 330–349. Bibcode:1986QuRes..25..330D. doi:10.1016/0033-5894(86)90005-0

Various Subdisciplines of Botany

Botany can be divided into various subdisciplines like agronomy, ethnobotany, paleobotany and bryology. Plant morphology studies the external parts of a plant whereas agronomy studies the plants used for the purpose of food, fiber and fuel. This chapter provides a plethora of interdisciplinary topics for better comprehension of botany.

Plant Morphology

Plant morphology or phytomorphology is the study of the physical form and external structure of plants. This is usually considered distinct from plant anatomy, which is the study of the internal structure of plants, especially at the microscopic level. Plant morphology is useful in the visual identification of plants.

Inflorescences emerging from protective coverings

Scope

Plant morphology "represents a study of the development, form, and structure of plants, and, by implication, an attempt to interpret these on the basis of similarity of plan and origin." There are four major areas of investigation in plant morphology, and each overlaps with another field of the biological sciences.

First of all, morphology is comparative, meaning that the morphologist examines structures in many different plants of the same or different species, then draws comparisons and formulates ideas about similarities. When structures in different species are believed to exist and develop as a result of common, inherited genetic pathways, those structures are termed homologous. For example, the leaves of pine, oak, and cabbage all look very different, but share certain basic structures and arrangement of parts. The homology of leaves is an easy conclusion to make. The plant morphologist goes further, and discovers that the spines of cactus also share the same basic structure and development as leaves in other plants, and therefore cactus spines are homologous to leaves as well. This aspect of plant morphology overlaps with the study of plant evolution and paleobotany.

Asclepias syriaca showing complex morphology of the flowers

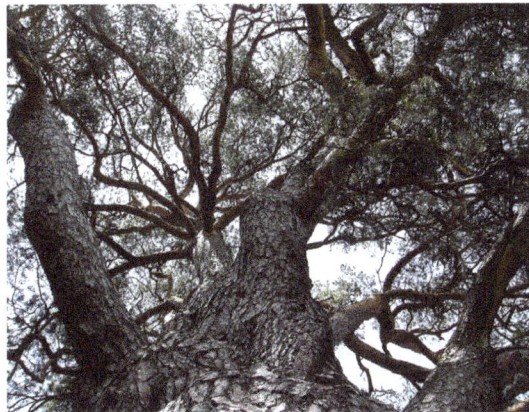

Looking up into the branch structure of a *Pinus sylvestris* tree

Secondly, plant morphology observes both the vegetative (somatic) structures of plants, as well as the reproductive structures. The vegetative structures of vascular plants includes the study of the shoot system, composed of stems and leaves, as well as the root system. The reproductive structures are more varied, and are usually specific to a particular group of plants, such as flowers and seeds, fern sori, and moss capsules. The detailed study of reproductive structures in plants led to the discovery of the alternation of generations found in all plants and most algae. This area of plant morphology overlaps with the study of biodiversity and plant systematics.

Thirdly, plant morphology studies plant structure at a range of scales. At the smallest scales are ultrastructure, the general structural features of cells visible only with the aid of an electron microscope, and cytology, the study of cells using optical microscopy. At this scale, plant morphology overlaps with plant anatomy as a field of study. At the largest scale is the study of plant growth habit, the overall architecture of a plant. The pattern of branching in a tree will vary from species to species, as will the appearance of a plant as a tree, herb, or grass.

Fourthly, plant morphology examines the pattern of development, the process by which structures originate and mature as a plant grows. While animals produce all the body parts they will ever have from early in their life, plants constantly produce new tissues and structures throughout their life. A living plant always has embryonic tissues. The way in which new structures mature as they are produced may be affected by the point in the plant's life when they begin to develop, as well as by the environment to which the structures are exposed. A morphologist studies this process, the causes, and its result. This area of plant morphology overlaps with plant physiology and ecology.

A Comparative Science

A plant morphologist makes comparisons between structures in many different plants of the same or different species. Making such comparisons between similar structures in different plants tackles the question of *why* the structures are similar. It is quite likely that similar underlying causes of genetics, physiology, or response to the environment have led to this similarity in appearance. The result of scientific investigation into these causes can lead to one of two insights into the underlying biology:

1. Homology - the structure is similar between the two species because of shared ancestry and common genetics.

2. Convergence - the structure is similar between the two species because of independent adaptation to common environmental pressures.

Understanding which characteristics and structures belong to each type is an important part of understanding plant evolution. The evolutionary biologist relies on the plant morphologist to interpret structures, and in turn provides phylogenies of plant relationships that may lead to new morphological insights.

Homology

When structures in different species are believed to exist and develop as a result of common, inherited genetic pathways, those structures are termed *homologous*. For example, the leaves of pine, oak, and cabbage all look very different, but share certain basic structures and arrangement of parts. The homology of leaves is an easy conclusion to make. The plant morphologist goes further, and discovers that the spines of cactus also share the same basic structure and development as leaves in other plants, and therefore cactus spines are homologous to leaves as well.

Convergence

When structures in different species are believed to exist and develop as a result of common adaptive responses to environmental pressure, those structures are termed *convergent*. For example, the fronds of *Bryopsis plumosa* and stems of *Asparagus setaceus* both have the same feathery branching appearance, even though one is an alga and one is a flowering plant. The similarity in overall structure occurs independently as a result of convergence. The growth form of many cacti and species of *Euphorbia* is very similar, even though they belong to widely distant families. The similarity results from common solutions to the problem of surviving in a hot, dry environment.

Euphorbia obesa, a spurge

Astrophytum asterias, a cactus

Vegetative and Reproductive Characteristics

Plant morphology treats both the vegetative structures of plants, as well as the reproductive structures.

The vegetative (somatic) structures of vascular plants include two major organ systems: (1) a shoot system, composed of stems and leaves, and (2) a root system. These two systems are common to nearly all vascular plants, and provide a unifying theme for the study of plant morphology.

By contrast, the reproductive structures are varied, and are usually specific to a particular group of plants. Structures such as flowers and fruits are only found in the angiosperms; sori are only found in ferns; and seed cones are only found in conifers and other gymnosperms. Reproductive characters are therefore regarded as more useful for the classification of plants than vegetative characters.

Use in Identification

Plant biologists use morphological characters of plants which can be compared, measured, counted and described to assess the differences or similarities in plant taxa and use these characters for plant identification, classification and descriptions.

When characters are used in descriptions or for identification they are called diagnostic or key characters which can be either qualitative and quantitative.

1. Quantitative characters are morphological features that can be counted or measured for example a plant species has flower petals 10–12 mm wide.

2. Qualitative characters are morphological features such as leaf shape, flower color or pubescence.

Both kinds of characters can be very useful for the identification of plants.

Alternation of Generations

The detailed study of reproductive structures in plants led to the discovery of the alternation of generations, found in all plants and most algae, by the German botanist Wilhelm Hofmeister. This discovery is one of the most important made in all of plant morphology, since it provides a common basis for understanding the life cycle of all plants.

Pigmentation in Plants

The primary function of pigments in plants is photosynthesis, which uses the green pigment chlorophyll along with several red and yellow pigments that help to capture as much light energy as possible. Pigments are also an important factor in attracting insects to flowers to encourage pollination.

Plant pigments include a variety of different kinds of molecule, including porphyrins, carotenoids, anthocyanins and betalains. All biological pigments selectively absorb certain wavelengths of light while reflecting others. The light that is absorbed may be used by the plant to power chemical reactions, while the reflected wavelengths of light determine the color the pigment will appear to the eye.

Morphology in Development

Plant development is the process by which structures originate and mature as a plant grows. It is a subject studies in plant anatomy and plant physiology as well as plant morphology.

The process of development in plants is fundamentally different from that seen in vertebrate animals. When an animal embryo begins to develop, it will very early produce all of the body parts that it will ever have in its life. When the animal is born (or hatches from its egg), it has all its body parts and from that point will only grow larger and more mature. By contrast, plants constantly produce new tissues and structures throughout their life from meristems located at the tips of organs, or between mature tissues. Thus, a living plant always has embryonic tissues.

The properties of organization seen in a plant are emergent properties which are more than the sum of the individual parts. "The assembly of these tissues and functions into an integrated multicellular organism yields not only the characteristics of the separate parts and processes but also quite a new set of characteristics which would not have been predictable on the basis of examination of the separate parts." In other words, knowing everything about the molecules in a plant are not enough to predict characteristics of the cells; and knowing all the properties of the cells will not predict all the properties of a plant's structure.

Growth

A vascular plant begins from a single celled zygote, formed by fertilisation of an egg cell by a sperm cell. From that point, it begins to divide to form a plant embryo through the process of embryogenesis. As this happens, the resulting cells will organize so that one end becomes the first root, while the other end forms the tip of the shoot. In seed plants, the embryo will develop one or more "seed leaves" (cotyledons). By the end of embryogenesis, the young plant will have all the parts necessary to begin in its life.

Once the embryo germinates from its seed or parent plant, it begins to produce additional organs (leaves, stems, and roots) through the process of organogenesis. New roots grow from root meristems located at the tip of the root, and new stems and leaves grow from shoot meristems located at the tip of the shoot. Branching occurs when small clumps of cells left behind by the meristem, and which have not yet undergone cellular differentiation to form a specialized tissue, begin to grow as the tip of a new root or shoot. Growth from any such meristem at the tip of a root or shoot is termed primary growth and results in the lengthening of that root or shoot. Secondary growth results in widening of a root or shoot from divisions of cells in a cambium.

In addition to growth by cell division, a plant may grow through cell elongation. This occurs when individual cells or groups of cells grow longer. Not all plant cells will grow to the same length. When cells on one side of a stem grow longer and faster than cells on the other side, the stem will bend to the side of the slower growing cells as a result. This directional growth can occur via a plant's response to a particular stimulus, such as light (phototropism), gravity (gravitropism), water, (hydrotropism), and physical contact (thigmotropism).

Plant growth and development are mediated by specific plant hormones and plant growth regulators (PGRs) (Ross et al. 1983). Endogenous hormone levels are influenced by plant age, cold hardiness, dormancy, and other metabolic conditions; photoperiod, drought, temperature, and other external environmental conditions; and exogenous sources of PGRs, e.g., externally applied and of rhizospheric origin.

Morphological Variation

Plants exhibit natural variation in their form and structure. While all organisms vary from individual to individual, plants exhibit an additional type of variation. Within a single individual, parts are repeated which may differ in form and structure from other similar parts. This variation is most easily seen in the leaves of a plant, though other organs such as stems and flowers may show similar variation. There are three primary causes of this variation: positional effects, environmental effects, and juvenility.

Evolution of Plant Morphology

Transcription factors and transcriptional regulatory networks play key roles in plant morphogenesis and their evolution. During plant landing, many novel transcription factor families emerged and are preferentially wired into the networks of multicellular development, reproduction, and organ development, contributing to more complex morphogenesis of land plants.

Positional Effects

Variation in leaves from the giant ragweed illustrating positional effects. The lobed leaves come from the base of the plant, while the unlobed leaves come from the top of the plant

Although plants produce numerous copies of the same organ during their lives, not all copies of a particular organ will be identical. There is variation among the parts of a mature plant resulting from the relative position where the organ is produced. For example, along a new branch the leaves may vary in a consistent pattern along the branch. The form of leaves produced near the base of the branch will differ from leaves produced at the tip of the plant, and this difference is consistent from branch to branch on a given plant and in a given species. This difference persists after the leaves at both ends of the branch have matured, and is not the result of some leaves being younger than others.

Environmental Effects

The way in which new structures mature as they are produced may be affected by the point in the plants life when they begin to develop, as well as by the environment to which the structures are exposed. This can be seen in aquatic plants and emergent plants.

Temperature

Temperature has a multiplicity of effects on plants depending on a variety of factors, including the size and condition of the plant and the temperature and duration of exposure. The smaller and more succulent the plant, the greater the susceptibility to damage or death from temperatures that are too high or too low. Temperature affects the rate of biochemical and physiological processes, rates generally (within limits) increasing with temperature. However, the Van't Hoff relationship for monomolecular reactions (which states that the velocity of a reaction is doubled or trebled by a temperature increase of $10°$ C) does not strictly hold for biological processes, especially at low and high temperatures.

When water freezes in plants, the consequences for the plant depend very much on whether the freezing occurs intracellularly (within cells) or outside cells in intercellular (extracellular) spaces. Intracellular freezing usually kills the cell regardless of the hardiness of the plant and its tissues. Intracellular freezing seldom occurs in nature, but moderate rates of decrease in temperature, e.g., $1°$ C to $6°$ C/hour, cause intercellular ice to form, and this "extraorgan ice" may or may not be lethal, depending on the hardiness of the tissue.

At freezing temperatures, water in the intercellular spaces of plant tissues freezes first, though the water may remain unfrozen until temperatures fall below 7° C. After the initial formation of ice intercellularly, the cells shrink as water is lost to the segregated ice. The cells undergo freeze-drying, the dehydration being the basic cause of freezing injury.

The rate of cooling has been shown to influence the frost resistance of tissues, but the actual rate of freezing will depend not only on the cooling rate, but also on the degree of supercooling and the properties of the tissue. Sakai (1979a) demonstrated ice segregation in shoot primordia of Alaskan white and black spruces when cooled slowly to 30° C to -40° C. These freeze-dehydrated buds survived immersion in liquid nitrogen when slowly rewarmed. Floral primordia responded similarly. Extraorgan freezing in the primordia accounts for the ability of the hardiest of the boreal conifers to survive winters in regions when air temperatures often fall to -50° C or lower. The hardiness of the winter buds of such conifers is enhanced by the smallness of the buds, by the evolution of faster translocation of water, and an ability to tolerate intensive freeze dehydration. In boreal species of *Picea* and *Pinus*, the frost resistance of 1-year-old seedlings is on a par with mature plants, given similar states of dormancy.

Juvenility

Juvenility in a seedling of European beech. There is a marked difference in shape between the first dark green "seed leaves" and the lighter second pair of leaves

The organs and tissues produced by a young plant, such as a seedling, are often different from those that are produced by the same plant when it is older. This phenomenon is known as juvenility or heteroblasty. For example, young trees will produce longer, leaner branches that grow upwards more than the branches they will produce as a fully grown tree. In addition, leaves produced during early growth tend to be larger, thinner, and more irregular than leaves on the adult plant. Specimens of juvenile plants may look so completely different from adult plants of the same species that egg-laying insects do not recognize the plant as food for their young. Differences are seen in rootability and flowering and can be seen in the same mature tree. Juvenile cuttings taken from the base of a tree will form roots much more readily than cuttings originating from the mid to upper crown. Flowering close to the base of a tree is absent or less profuse than flowering in the higher branches especially when a young tree first reaches flowering age.

The transition from early to late growth forms is referred to as 'vegetative phase change', but there is some disagreement about terminology.

Modern Morphology

Rolf Sattler has revised fundamental concepts of comparative morphology such as the concept of homology. He emphasized that homology should also include partial homology and quantitative homology. This leads to a continuum morphology that demonstrates a continuum between the morphological categories of root, shoot, stem (caulome), leaf (phyllome), and hair (trichome). How intermediates between the categories are best described has been discussed by Bruce K. Kirchoff et al.

Honoring Agnes Arber, author of the partial-shoot theory of the leaf, Rutishauser and Isler called the continuum approach Fuzzy Arberian Morphology (FAM). "Fuzzy" refers to fuzzy logic, "Arberian" to Agnes Arber. Rutishauser and Isler emphasized that this approach is not only supported by many morphological data but also by evidence from molecular genetics. More recent evidence from molecular genetics provides further support for continuum morphology. James (2009) concluded that "it is now widely accepted that... radiality [characteristic of most shoots] and dorsiventrality [characteristic of leaves] are but extremes of a continuous spectrum. In fact, it is simply the timing of the KNOX gene expression!." Eckardt and Baum (2010) concluded that "it is now generally accepted that compound leaves express both leaf and shoot properties."

Process morphology (dynamic morphology) describes and analyzes the dynamic continuum of plant form. According to this approach, structures do not *have* process(es), they *are* process(es). Thus, the structure/process dichotomy is overcome by "an enlargement of our concept of 'structure' so as to include and recognize that in the living organism it is not merely a question of spatial structure with an 'activity' as something over or against it, but that the concrete organism is a spatio-temporal structure and that this spatio-temporal structure is the activity itself."

For Jeune, Barabé and Lacroix, classical morphology (that is, mainstream morphology, based on a qualitative homology concept implying mutually exclusive categories) and continuum morphology are sub-classes of the more encompassing process morphology (dynamic morphology).

Classical morphology, continuum morphology, and process morphology are highly relevant to plant evolution, especially the field of plant evolutionary biology (plant evo-devo) that tries to integrate plant morphology and plant molecular genetics. In a detailed case study on unusual morphologies, Rutishauser (2016) illustrated and discussed various topics of plant evo-devo such as the fuzziness (continuity) of morphological concepts, the lack of a one-to-one correspondence between structural categories and gene expression, the notion of morphospace, the adaptive value of bauplan features versus patio ludens, physiological adaptations, hopeful monsters and saltational evolution, the significance and limits of developmental robustness, etc.

Agronomy

Agronomy is the science and technology of producing and using plants for food, fuel, fiber, and land reclamation. Agronomy has come to encompass work in the areas of plant genetics, plant physiology, meteorology, and soil science. It is the application of a combination of sciences like biology, chemistry, economics, ecology, earth science, and genetics. Agronomists of today are involved with many issues, including producing food, creating healthier food, managing the environmental impact of agriculture, and extracting energy from plants. Agronomists often specialise in areas such as crop rotation, irrigation and drainage, plant breeding, plant physiology, soil classification, soil fertility, weed control, and insect and pest control.

Plant Breeding

An agronomist field sampling a trial plot of flax

This area of agronomy involves selective breeding of plants to produce the best crops under various conditions. Plant breeding has increased crop yields and has improved the nutritional value of numerous crops, including corn, soybeans, and wheat. It has also led to the development of new types of plants. For example, a hybrid grain called triticale was produced by crossbreeding rye and wheat. Triticale contains more usable protein than does either rye or wheat. Agronomy has also been instrumental in fruit and vegetable production research.

Biotechnology

Purdue University agronomy professor George Van Scoyoc explains the difference between forest and prairie soils to soldiers of the Indiana National Guard's Agribusiness Development Team at the Beck Agricultural Center in West Lafayette, Indiana

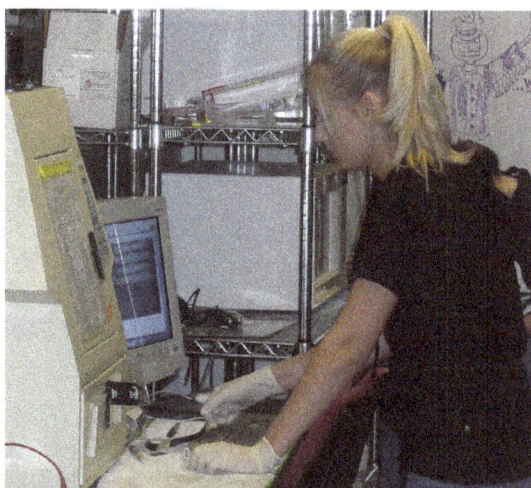

An agronomist mapping a plant genome

Agronomists use biotechnology to extend and expedite the development of desired characteristic. Biotechnology is often a lab activity requiring field testing of the new crop varieties that are developed.

In addition to increasing crop yields agronomic biotechnology is increasingly being applied for novel uses other than food. For example, oilseed is at present used mainly for margarine and other food oils, but it can be modified to produce fatty acids for detergents, substitute fuels and petrochemicals.

Soil Science

Agronomists study sustainable ways to make soils more productive and profitable. They classify soils and analyze them to determine whether they contain nutrients vital to plant growth. Common macronutrients analyzed include compounds of nitrogen, phosphorus, potassium, calcium, magnesium, and sulfur. Soil is also assessed for several micronutrients, like zinc and boron. The percentage of organic matter, soil pH, and nutrient holding capacity (cation exchange capacity) are tested in a regional laboratory. Agronomists will interpret these lab reports and make recommendations to balance soil nutrients for optimal plant growth.

Soil Conservation

In addition, agronomists develop methods to preserve the soil and to decrease the effects of erosion by wind and water. For example, a technique called contour plowing may be used to prevent soil erosion and conserve rainfall. Researchers in agronomy also seek ways to use the soil more effectively in solving other problems. Such problems include the disposal of human and animal manure, water pollution, and pesticide build-up in the soil. Techniques include no-tilling crops, planting of soil-binding grasses along contours on steep slopes, and contour drains of depths up to 1 metre.

Agroecology

Agroecology is the management of agricultural systems with an emphasis on ecological and environmental perspectives. This area is closely associated with work in the areas of sustainable agriculture, organic farming, and alternative food systems and the development of alternative cropping systems.

Theoretical Modeling

Theoretical production ecology tries to quantitatively study the growth of crops. The plant is treated as a kind of biological factory, which processes light, carbon dioxide, water, and nutrients into harvestable parts. Main parameters kept into consideration are temperature, sunlight, standing crop biomass, plant production distribution, nutrient and water supply.

Ethnobotany

The ethnobotanist Richard Evans Schultes at work in the Amazon

Ethnobotany is the study of a region's plants and their practical uses through the traditional knowledge of a local culture and people. An ethnobotanist thus strives to document the local customs involving the practical uses of local flora for many aspects of life, such as plants as medicines, foods, and clothing. Richard Evans Schultes, often referred to as the "father of ethnobotany", explained the discipline in this way:

> "Ethnobotany simply means [...] investigating plants used by primitive societies in various parts of the world."

Since the time of Schultes, the field of ethnobotany has grown from simply acquiring ethnobotanical knowledge to that of applying it to a modern society, primarily in the form of pharmaceuticals. Intellectual property rights and benefit-sharing arrangements are important issues in ethnobotany.

History

Plants have been widely used by American Indian healers, such as this Ojibwa man

The idea of ethnobotany was first proposed by the early 20th century botanist John William Harshberger. While Harshberger did perform ethnobotanical research extensively, including in areas such as North Africa, Mexico, Scandinavia, and Pennsylvania, it was not until Richard Evans Schultes began his trips into the Amazon that ethnobotany become a more well known science. However, the practice of ethnobotany is thought to have much earlier origins in the first century AD when a Greek physician by the name of Pedanius Dioscorides wrote an extensive botanical text detailing the medical and cullinary properties of "over 600 mediteranian plants" named De Materia Medica. Historians note that Dioscorides wrote about traveling often throughout the Roman empire, including regions such as "Greece, Crete, Egypt, and Petra", and in doing so obtained substantial knowledge about the local plants and their useful properties. European botanical knowledge drastically expanded once the New World was discovered due to ethnobotany. This expansion in knowledge can be primarily attributed to the substantial influx of new plants from the Americas, including crops such as potatoes, peanuts, avocados, and tomatoes. One French explorer in the 16th century, Jacques Cartier, discovered a cure for scurvy (a tea made from boiling the bark of the Sitka Spruce) from a local Iroquoi tribe.

Medieval and Renaissance

During the medieval period, ethnobotanical studies were commonly found connected with monasticism. Notable at this time was Hildegard von Bingen. However, most botanical knowledge was kept in gardens such as physic gardens attached to hospitals and religious buildings. It was thought of in practical use terms for culinary and medical purposes and the ethnographic element was not studied as a modern anthropologist might approach ethnobotany today.

The Age of Reason

Carl Linnaeus carried out in 1732 a research expedition in Scandinavia asking the Sami people about their ethnological usage of plants.

The age of enlightenment saw a rise in economic botanical exploration. Alexander von Humboldt collected data from the New World, and James Cook's voyages brought back collections and information on plants from the South Pacific. At this time major botanical gardens were started, for instance the Royal Botanic Gardens, Kew in 1759. The directors of the gardens sent out gardener-botanist explorers to care for and collect plants to add to their collections.

As the 18th century became the 19th, ethnobotany saw expeditions undertaken with more colonial aims rather than trade economics such as that of Lewis and Clarke which recorded both plants and the peoples encountered use of them. Edward Palmer collected material culture artifacts and botanical specimens from people in the North American West (Great Basin) and Mexico from the 1860s to the 1890s. Through all of this research, the field of "aboriginal botany" was established—the study of all forms of the vegetable world which aboriginal peoples use for food, medicine, textiles, ornaments and more.

As a Modern Science

The first individual to study the emic perspective of the plant world was a German physician working in Sarajevo at the end of 19th century: Leopold Glueck. His published work on traditional medical uses of plants done by rural people in Bosnia (1896) has to be considered the first modern ethnobotanical work.

Other scholars analyzed uses of plants under an indigenous/local perspective in the 20th century: Matilda Coxe Stevenson, Zuni plants (1915); Frank Cushing, Zuni foods (1920); Keewaydinoquay Peschel, Anishinaabe fungi (1998), and the team approach of Wilfred Robbins, John Peabody Harrington, and Barbara Freire-Marreco, Tewa pueblo plants (1916).

In the beginning, ethonobotanical specimens and studies were not very reliable and sometimes not helpful. This is because the botanists and the anthropologists did not always collaborate in their work. The botanists focused on identifying species and how the plants were used instead of concentrating upon how plants fit into people's lives. On the other hand, anthropologists were interested in the cultural role of plants and treated other scientific aspects superficially. In the early 20th century, botanists and anthropologists better collaborated and the collection of reliable, detailed cross-disciplinary data began.

Beginning in the 20th century, the field of ethnobotany experienced a shift from the raw compilation of data to a greater methodological and conceptual reorientation. This is also the beginning of academic ethnobotany. The so-called "father" of this discipline is Richard Evans Schultes, even though he did not actually coin the term "ethnobotany". Today the field of ethnobotany requires a variety of skills: botanical training for the identification and preservation of plant specimens; anthropological training to understand the cultural concepts around the perception of plants; linguistic training, at least enough to transcribe local terms and understand native morphology, syntax, and semantics.

Mark Plotkin, who studied at Harvard University, the Yale School of Forestry and Tufts University, has contributed a number of books on ethnobotany. He completed a handbook for the Tirio people of Suriname detailing their medicinal plants; *Tales of a Shaman's Apprentice* (1994); *The Shaman's Apprentice,* a children's book with Lynne Cherry (1998); and *Medicine Quest: In Search of Nature's Healing Secrets* (2000).

Plotkin was interviewed in 1998 by *South American Explorer* magazine, just after the release of *Tales of a Shaman's Apprentice* and the IMAX movie Amazonia. In the book, he stated that he saw wisdom in both traditional and Western forms of medicine:

No medical system has all the answers—no shaman that I've worked with has the equivalent of a polio vaccine and no dermatologist that I've been to could cure a fungal infection as effectively (and inexpensively) as some of my Amazonian mentors. It shouldn't be the doctor versus the witch doctor. It should be the best aspects of all medical systems (ayurvedic, herbalism, homeopathic, and so on) combined in a way which makes health care more effective and more affordable for all.

A great deal of information about the traditional uses of plants is still intact with tribal peoples. But the native healers are often reluctant to accurately share their knowledge to outsiders. Schultes actually apprenticed himself to an Amazonian shaman, which involves a long-term commitment and genuine relationship. In *Wind in the Blood: Mayan Healing & Chinese Medicine* by Garcia et al. the visiting acupuncturists were able to access levels of Mayan medicine that anthropologists could not because they had something to share in exchange. Cherokee medicine priest David Winston describes how his uncle would invent nonsense to satisfy visiting anthropologists.

Another scholar, James W. Herrick, who studied under ethnologist William N. Fenton, in his work *Iroquois Medical Ethnobotany* (1995) with Dean R. Snow (editor), professor of Anthropology at Penn State, explains that understanding herbal medicines in traditional Iroquois cultures is rooted in a strong and ancient cosmological belief system. Their work provides perceptions and conceptions of illness and imbalances which can manifest in physical forms from benign maladies to serious diseases. It also includes a large compilation of Herrick's field work from numerous Iroquois authorities of over 450 names, uses, and preparations of plants for various ailments. Traditional Iroquois practitioners had (and have) a sophisticated perspective on the plant world that contrast strikingly with that of modern medical science.

Issues

Many instances of gender bias have occurred in ethnobotany, creating the risk of drawing erroneous conclusions. Other issues include ethical concerns regarding interactions with indigenous populations, and the International Society of Ethnobiology has created a code of ethics to guide researchers.

Bryology

Bryology (from Greek *bryon*, a moss, a liverwort) is the branch of botany concerned with the scientific study of bryophytes (mosses, liverworts, and hornworts). Bryologists are people who have an active interest in observing, recording, classifying or researching bryophytes. The field is often studied along with lichenology due to the similar appearance and ecological niche of the two organisms, even though bryophytes and lichens are not classified in the same kingdom.

Common bryophytes found in central Japan

History

Bryophytes were first studied in detail in the 18th century. The German botanist Johann Jacob Dillenius (1687–1747) was a professor at Oxford and in 1717 produced the work "Reproduction of the ferns and mosses." The beginning of bryology really belongs to the work of Johannes Hedwig, who clarified the reproductive system of mosses (1792, *Fundamentum historiae naturalist muscorum*) and arranged a taxonomy.

Areas of research include bryophyte taxonomy, bryophytes as bioindicators, DNA sequencing, and the interdependency of bryophytes and other plant and animal species. Among other things, scientists have discovered parasitic bryophytes such as Cryptothallus and potentially carnivorous liverworts such as Colura zoophaga and Pleurozia.

Centers of research in bryology include University of Bonn, Germany, the University of Helsinki, Finland and the New York Botanical Garden.

Paleobotany

Paleobotany, also spelled as palaeobotany (from the Greek words *paleon* = old and "botany", study of plants), is the branch of paleontology or paleobiology dealing with the recovery and identification of plant remains from geological contexts, and their use for the biological reconstruction of past environments (paleogeography), and both the evolutionary history of plants, with a bearing upon the evolution of life in general. A synonym is paleophytology. Paleobotany includes the study of terrestrial plant fossils, as well as the study of prehistoric marine photoautotrophs, such as photosynthetic algae, seaweeds or kelp. A closely related field is palynology, which is the study of fossilized and extant spores and pollen.

Paleobotany is important in the reconstruction of ancient ecological systems and climate, known as paleoecology and paleoclimatology respectively; and is fundamental to the study of green plant development and evolution. Paleobotany has also become important to the field of archaeology, primarily for the use of phytoliths in relative dating and in paleoethnobotany.

A leaf fossil of the European beech (*Fagus sylvatica*) from the late Pliocene of France, approximately 3 million years ago

The emergence of paleobotany as a scientific discipline can be seen in the early 19th century, especially in the works of the German palaeontologist Ernst Friedrich von Schlotheim, the Czech (Bohemian) nobleman and scholar Kaspar Maria von Sternberg, and the French botanist Adolphe-Théodore Brongniart.

Overview of the Paleobotanical Record

Macroscopic remains of true vascular plants are first found in the fossil record during the Silurian Period of the Paleozoic era. Some dispersed, fragmentary fossils of disputed affinity, primarily spores and cuticles, have been found in rocks from the Ordovician Period in Oman, and are thought to derive from liverwort- or moss-grade fossil plants (Wellman, Osterloff & Mohiuddin 2003).

An important early land plant fossil locality is the Rhynie Chert, found outside the village of Rhynie in Scotland. The Rhynie chert is an Early Devonian sinter (hot spring) deposit composed primarily of silica. It is exceptional due to its preservation of several different clades of plants, from mosses and lycopods to more unusual, problematic forms. Many fossil animals, including arthropods and arachnids, are also found in the Rhynie Chert, and it offers a unique window on the history of early terrestrial life.

An unpolished hand sample of the Lower Devonian Rhynie Chert from Scotland

Plant-derived macrofossils become abundant in the Late Devonian and include tree trunks, fronds, and roots. The earliest tree was thought to be *Archaeopteris*, which bears simple, fern-like leaves spirally arranged on branches atop a conifer-like trunk (Meyer-Berthaud, Scheckler & Wendt 1999), though it is now known to be the recently discovered *Wattieza*.

Widespread coal swamp deposits across North America and Europe during the Carboniferous Period contain a wealth of fossils containing arborescent lycopods up to 30 meters tall, abundant seed plants, such as conifers and seed ferns, and countless smaller, herbaceous plants.

Angiosperms (flowering plants) evolved during the Mesozoic, and flowering plant pollen and leaves first appear during the Early Cretaceous, approximately 130 million years ago.

Plant Fossils

A plant fossil is any preserved part of a plant that has long since died. Such fossils may be prehistoric impressions that are many millions of years old, or bits of charcoal that are only a few hundred years old. Prehistoric plants are various groups of plants that lived before recorded history (before about 3500 BC).

Preservation of Plant Fossils

Ginkgoites huttonii, Middle Jurassic, Yorkshire, UK. Leaves preserved as compressions.
Specimen in Munich Palaeontological Museum, Germany.

Plant fossils can be preserved in a variety of ways, each of which can give different types of information about the original parent plant. These modes of preservation are discussed in the general pages on fossils but may be summarised in a palaeobotanical context as follows.

1. Adpressions (compressions - impressions). These are the most commonly found type of plant fossil. They provide good morphological detail, especially of dorsiventral (flattened) plant parts such as leaves. If the cuticle is preserved, they can also yield fine anatomical detail of the epidermis. Little other detail of cellular anatomy is normally preserved.

2. Petrifactions (permineralisations or anatomically preserved fossils). These provide fine detail of the cell anatomy of the plant tissue. Morphological detail can also be determined by serial sectioning, but this is both time consuming and difficult.

3. Moulds and casts. These only tend to preserve the more robust plant parts such as seeds or woody stems. They can provide information about the three-dimensional form of the plant, and in the case of casts of tree stumps can provide evidence of the density of the original vegetation. However, they rarely preserve any fine morphological detail or cell anatomy. A subset of such fossils are pith casts, where the centre of a stem is either hollow or has delicate pith. After death, sediment enters and forms a cast of the central cavity of the stem. The best known examples of pith casts are in the Carboniferous Sphenophyta (*Calamites*) and cordaites (*Artisia*).

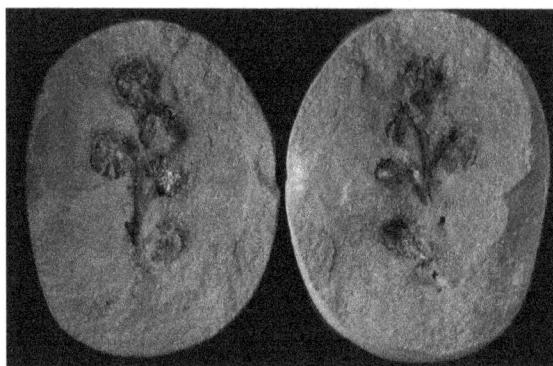

Crossotheca hughesiana Kidston, Middle Pennsylvanian, Coseley, near Dudley, UK. A lyginopteridalean pollen organ preserved as an authigenic mineralization (mineralized *in situ*). Specimen in Sedgwick Museum, Cambridge, UK.

4. Authigenic mineralisations. These can provide very fine, three-dimensional morphological detail, and have proved especially important in the study of reproductive structures that can be severely distorted in adpressions. However, as they are formed in mineral nodules, such fossils can rarely be of large size.

5. Fusain. Fire normally destroys plant tissue but sometimes charcoalified remains can preserve fine morphological detail that is lost in other modes of preservation; some of the best evidence of early flowers has been preserved in fusain. Fusain fossils are delicate and often small, but because of their buoyancy can often drift for long distances and can thus provide evidence of vegetation away from areas of sedimentation.

Fossil-taxa

Plant fossils almost always represent disarticulated parts of plants; even small herbaceous plants are rarely preserved whole. Those few examples of plant fossils that appear to be the remains of whole plants in fact are incomplete as the internal cellular tissue and fine micromorphological detail is normally lost during fossilisation. Plant remains can be preserved in a variety of ways, each revealing different features of the original parent plant.

Because of these difficulties, palaeobotanists usually assign different taxonomic names to different parts of the plant in different modes of preservation. For instance, in the subarborescent Palaeozoic sphenophytes, an impression of a leaf might be assigned to the genus *Annularia*, a compression of a cone assigned to *Palaeostachya*, and the stem assigned to either *Calamites* or *Arthroxylon* depending on whether it is preserved as a cast or a petrifaction. All of these fossils may have originated from the same parent plant but they are each given their own taxonomic name. This approach to naming plant fossils originated with the work of Alexandre Brongniart and has stood the test of time.

For many years this approach to naming plant fossils was accepted by palaeobotanists but not formalised within the *International Rules of Botanical Nomenclature*. Eventually, Thomas (1935) and Jongmans, Halle & Gothan (1935) proposed a set of formal provisions, the essence of which was introduced into the 1952 International Code of Botanical Nomenclature. These early provisions allowed fossils representing particular parts of plants in a particular state of preservation to be referred to organ-genera. In addition, a small subset of organ-genera, to be known as form-genera, were recognised based on the artificial taxa introduced by Brongniart (1822) mainly for foliage fossils. Over the years, the concepts and regulations surrounding organ- and form-genera became modified within successive codes of nomenclature, reflecting a failure of the palaeobotanical community to agree on how this aspect of plant taxonomic nomenclature should work (A history reviewed by Cleal & Thomas (2010)). The use of organ- and fossil-genera was abandoned with the *St Louis Code* (Greuter et al. 2000), replaced by "morphotaxa".

The situation in the *Vienna Code* of 2005 was that any plant taxon whose type is a fossil, except Diatoms, can be described as a morphotaxon, a particular part of a plant preserved in a particular way. Although the name is always fixed to the type specimen, the circumscription (i.e. range of specimens that may be included within the taxon) is defined by the taxonomist who uses the name. Such a change in circumscription could result in an expansion of the range of plant parts and/or preservation states that can be incorporated within the taxon. For instance, a fossil-genus originally based on compressions of ovules could be used to include the multi-ovulate cupules within which the ovules were originally borne. A complication can arise if, in this case, there was an already named fossil-genus for these cupules. If palaeobotanists were confident that the type of the ovule fossil-genus and of the cupule fossil-genus could be included in the same genus, then the two names would compete as to being the correct one for the newly emended genus.

Morphotaxa were introduced to try to overcome the issue of competing names that represented different plant parts and/or preservation states. What would you do if the species-name of a pollen-organ was pre-dated by the species name of the type of pollen produced by that pollen organ. It was argued that palaeobotanists would be unhappy if the pollen organs were named using the taxonomic name whose type specimen is a pollen grain. As pointed out by Cleal & Thomas (2010), however, the risk of the name of a pollen grain supplanting the name of a pollen organ is most unlikely. Palaeobotanists would have to be totally confident that the type specimen of the pollen species, which would normally be a dispersed grain, definitely came from the same plant that produced the pollen organ. We know from modern plants that closely related but distinct species can produce virtually indistinguishable pollen. It would seem that morphotaxa offer no real advantage to palaeobotanists over normal fossil-taxa and the concept was abandoned with the 2011 botanical congress and the 2012 International Code of Nomenclature for algae, fungi, and plants.

Fossil Groups of Plants

Some plants have remained remarkably unchanged throughout earth's geological time scale. Early ferns had developed by the Mississippian, conifers by the Pennsylvanian. Some plants of prehistory are the same ones around today and are thus living fossils, such as *Ginkgo biloba* and *Sciadopitys verticillata*. Other plants have changed radically, or have gone extinct entirely.

Stigmaria, a common fossil tree root. Upper Carboniferous of northeastern Ohio.

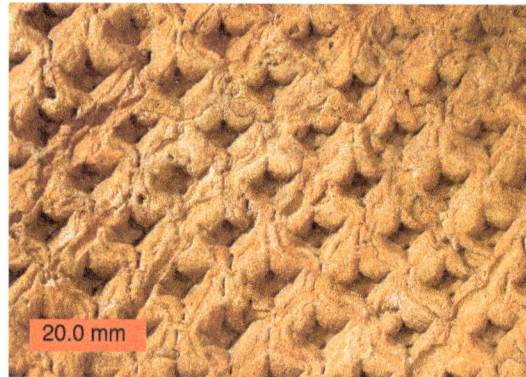

External mold of *Lepidodendron* from the Upper Carboniferous of Ohio.

Examples of prehistoric plants are:

- *Araucaria mirabilis*
- *Archaeopteris*
- *Calamites*
- *Dillhoffia*
- *Glossopteris*
- *Hymenaea protera*
- *Nelumbo aureavallis*
- *Pachypteris*
- *Palaeoraphe*
- *Peltandra primaeva*
- *Protosalvinia*
- *Trochodendron nastae*

References

- Raven, P. H., R. F. Evert, & S. E. Eichhorn. Biology of Plants, 7th ed., page 9. (New York: W. H. Freeman, 2005). ISBN 0-7167-1007-2

- Jones, Cynthia S. (1999-11-01). "An Essay on Juvenility, Phase Change, and Heteroblasty in Seed Plants". International Journal of Plant Sciences. 160 (S6): −105–S111. ISSN 1058-5893. doi:10.1086/314215. Retrieved 2016-10-16

- Barlow, P (2005). "Patterned cell determination in a plant tissue: The secondary phloem of trees". BioEssays. 27 (5): 533–41. PMID 15832381. doi:10.1002/bies.20214

- Sattler, Rolf (1992). "Process morphology: Structural dynamics in development and evolution". Canadian Journal of Botany. 70 (4): 708–714. doi:10.1139/b92-091

- Evert, Ray Franklin and Esau, Katherine (2006) Esau's Plant anatomy: meristems, cells, and tissues of the plant body - their structure, function and development Wiley, Hoboken, New Jersey, page xv, ISBN 0-471-73843-3

- Sattler, R. (1984). "Homology - a continuing challenge". Systematic Botany. 9 (4): 382–394. JSTOR 2418787. doi:10.2307/2418787

- Soejarto, D.D. et. al. (2005): "Ethnobotany/ethnopharmacology and mass bioprospecting: Issues on intellectual property and benefit-sharing", Journal of Ethnopharmacology, V. 100, 15-22

- Harold C. Bold, C. J. Alexopoulos, and T. Delevoryas. Morphology of Plants and Fungi, 5th ed., page 3. (New York: Harper-Collins, 1987). ISBN 0-06-040839-1

- "Brongniart, Adolphe-Théodore". www.encyclopedia.com. Encyclopedia.com: FREE online dictionary. Retrieved 22 February 2017

- Jeanine, M. Pfeiffer; Ramona, J. Butz. "Assessing Cultural And Ecological Variation In Ethnobiological Research: The Importance Of Gender" (PDF). Journal of Ethnobiology. 25 (2): 240–278. doi:10.2993/0278-0771(2005)25[240:ACAEVI]2.0.CO;2

- Vergara-Silva, Francisco (2003). "Plants and the Conceptual Articulation of Evolutionary Developmental Biology". Biology and Philosophy. 18 (2): 261–264. doi:10.1023/A:1023936102602

- Ponman, Bussmann, Bruce E, Rainer W. (2012). Medicinal Plants and the Legacy of Richard E. Schultes (PDF). Missouri Botanical Garden. ISBN 0984841520

- Bäurle, I; Laux, T (2003). "Apical meristems: The plant's fountain of youth". BioEssays. 25 (10): 961–70. PMID 14505363. doi:10.1002/bies.10341

- Michael A Dirr; Charles W Heuser, jr. (2006). "2". The Reference Manual of Woody Plant Propagation (Second ed.). Varsity Press Inc. pp. 26, 28, 29. ISBN 0942375092

Classification of Plants

Plants can be classified into different groups such as embryophyte, bryophyte, etc. Embryophytes are the most common type of green plants found on the surface of the Earth. They include mosses, liverworts, ferns, gymnosperms and lycophytes. The major classifications of plants are dealt with great details in the chapter.

Embryophyte

The Embryophyta are the most familiar group of green plants that form vegetation on earth. Living embryophytes include hornworts, liverworts, mosses, ferns, lycophytes, gymnosperms and flowering plants, and emerged from Charophyte green algae. The Embryophyta are informally called land plants because they live primarily in terrestrial habitats, while the related green algae are primarily aquatic. All are complex multicellular eukaryotes with specialized reproductive organs. The name derives from their innovative characteristic of nurturing the young embryo sporophyte during the early stages of its multicellular development within the tissues of the parent gametophyte. With very few exceptions, embryophytes obtain their energy by photosynthesis, that is by using the energy of sunlight to synthesize their food from carbon dioxide and water.

Description

The evolutionary origins of the embryophytes are discussed further below, but they are believed to have evolved from within a group of complex green algae during the Paleozoic era (which started around 540 million years ago). Charales or the stoneworts may be the best living illustration of that developmental step. Embryophytes are primarily adapted for life on land, although some are secondarily aquatic. Accordingly, they are often called *land plants* or *terrestrial plants*.

On a microscopic level, the cells of embryophytes are broadly similar to those of green algae, but differ in that in cell division the daughter nuclei are separated by a phragmoplast. They are eukaryotic, with a cell wall composed of cellulose and plastids surrounded by two membranes. The latter include chloroplasts, which conduct photosynthesis and store food in the form of starch, and are characteristically pigmented with chlorophylls *a* and *b*, generally giving them a bright green color. Embryophyte cells also generally have an enlarged central vacuole enclosed by a vacuolar membrane or tonoplast, which maintains cell turgor and keeps the plant rigid.

In common with all groups of multicellular algae they have a life cycle which involves 'alternation of generations'. A multicellular generation with a single set of chromosomes – the haploid gametophyte – produces sperm and eggs which fuse and grow into a multicellular generation with twice the number of chromosomes – the diploid sporophyte. The mature sporophyte produces haploid spores which grow into a gametophyte, thus completing the cycle. Embryophytes have two features related to their reproductive cycles which distinguish them from all other plant lineages. Firstly, their gametophytes produce sperm and eggs in multicellular structures (called 'antheridia' and 'archegonia'), and fertilization of the ovum takes place within the archegonium rather than in the external environment. Secondly, and most importantly, the initial stage of development of the fertilized egg (the zygote) into a diploid multicellular sporophyte, take place within the archegonium where it is both protected and provided with nutrition. This second feature is the origin of the term 'embryophyte' – the fertilized egg develops into a protected embryo, rather than dispersing as a single cell. In the bryophytes the sporophyte remains dependent on the gametophyte, while in all other embryophytes the sporophyte generation is dominant and capable of independent existence.

Embryophytes also differ from algae by having metamers. Metamers are repeated units of development, in which each unit derives from a single cell, but the resulting product tissue or part is largely the same for each cell. The whole organism is thus constructed from similar, repeating parts or *metamers*. Accordingly, these plants are sometimes termed 'metaphytes' and classified as the group Metaphyta (but Haeckel's definition of Metaphyta places some algae in this group). In all land plants a disc-like structure called a phragmoplast forms where the cell will divide, a trait only found in the land plants in the streptophyte lineage, some species within their relatives Coleochaetales, Charales and Zygnematales, as well as within subaerial species of the algae order Trentepohliales, and appears to be essential in the adaptation towards a terrestrial life style.

Phylogeny and Classification

All green algae and land plants are now known to form a single evolutionary lineage or clade, one name for which is Viridiplantae (i.e. 'green plants'). According to several molecular clock estimates the Viridiplantae split 1,200 million years ago to 725 million years ago into two clades: chlorophytes and streptophytes. The chlorophytes are considerably more diverse (with around 700 genera) and were originally marine, although some groups have since spread into fresh water. The streptophyte algae (i.e. the streptophyte clade minus the land plants) are less diverse (with around 122 genera) and adapted to fresh water very early in their evolutionary history. They have not spread into marine environments (only a few stoneworts, which belong to this group, tolerate brackish water). Some time during the Ordovician period (which started around 490 million years ago) one or more streptophytes invaded the land and began the evolution of the embryophyte land plants.

Becker and Marin speculate that land plants evolved from streptophytes rather than any other group of algae because streptophytes were adapted to living in fresh water. This prepared them to tolerate a range of environmental conditions found on land. Fresh water living made them tolerant of exposure to rain; living in shallow pools required tolerance to temperature variation, high levels of ultra-violet light and seasonal dehydration.

Relationships between the groups making up Viridiplantae are still being elucidated. Views have changed considerably since 2000 and classifications have not yet caught up. However, the division between chlorophytes and streptophytes and the evolution of embryophytes from within the latter group, as shown in the cladogram below, are well established. Three approaches to classification are shown. Older classifications, as on the left, treated all green algae as a single division of the plant kingdom under the name Chlorophyta. Land plants were then placed in separate divisions.

All the streptophyte algae can be grouped into one paraphyletic taxon, as in the middle, allowing the embryophytes to form a taxon at the same level. Alternatively, the embryophytes can be sunk into a monophyletic taxon comprising all the streptophytes, as shown. A variety of names have been used for the different groups which result from these approaches; those used below are only one of a number of possibilities. The higher-level classification of the Viridiplantae varies considerably, resulting in widely different ranks being assigned to the embryophytes, from kingdom to class.

Diversity

Bryophytes

Most bryophytes, such as these mosses, produce stalked sporophytes from which their spores are released

Bryophytes consist of all non-vascular land plants (embryophytes without vascular tissue). All are relatively small and are usually confined to environments that are humid or at least seasonally moist. They are limited by their reliance on water needed to disperse their gametes, although only a few bryophytes are truly aquatic. Most species are tropical, but there are many arctic species as well. They may locally dominate the ground cover in tundra and Arctic–alpine habitats or the epiphyte flora in rain forest habitats.

The three living divisions are the mosses (Bryophyta), hornworts (Anthocerotophyta), and liverworts (Marchantiophyta). Originally, these three groups were included together as classes within the single division Bryophyta. However, they now are placed separately into three divisions since the bryophytes as a whole are known to be a paraphyletic (artificial) group instead of a single lineage. Instead, the three bryophyte groups form an evolutionary grade of those land plants that are not vascular. Some closely related green algae are also non-vascular, but are not considered "land plants."

- Marchantiophyta (liverworts)

- Bryophyta (mosses)

- Anthocerotophyta (hornworts)

Despite the fact that they are no longer classified as a single group, the bryophytes are still studied together because of their many biological similarities as non-vascular land plants. All three bryophyte groups share a haploid-dominant life cycle and unbranched sporophytes. These are traits that appear to be plesiotypic within the land plants, and thus were common to all early diverging lineages of plants on the land. The fact that the bryophytes have a life cycle in common is thus an artefact of being the oldest extant lineages of land plant, and not the result of close shared ancestry.

The bryophyte life-cycle is strongly dominated by the haploid gametophyte generation. The sporophyte remains small and dependent on the parent gametophyte for its entire brief life. All other living groups of land plants have a life cycle dominated by the diploid sporophyte generation. It is in the diploid sporophyte that vascular tissue develops. Although some mosses have quite complex water-conducting vessels, bryophytes lack true vascular tissue.

Like the vascular plants, bryophytes do have differentiated stems, and although these are most often no more than a few centimeters tall, they do provide mechanical support. Most bryophytes also have leaves, although these typically are one cell thick and lack veins. Unlike the vascular plants, bryophytes lack true roots or any deep anchoring structures. Some species do grow a filamentous network of horizontal stems, but these have a primary function of mechanical attachment rather than extraction of soil nutrients (Palaeos 2008).

Rise of Vascular Plants

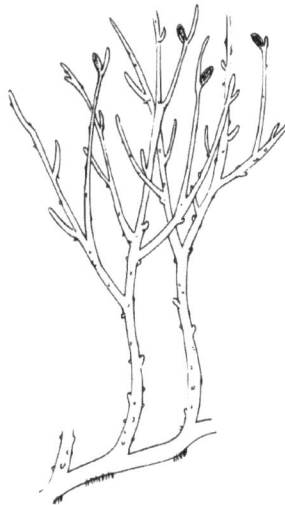

Reconstruction of a plant of *Rhynia*

During the Silurian and Devonian periods (around 440 to 360 million years ago), plants evolved which possessed true vascular tissue, including cells with walls strengthened by lignin (tracheids). Some extinct early plants appear to be between the grade of organization of bryophytes and that of true vascular plants (eutracheophytes). Genera such as *Horneophyton* have water-conducting tissue more like that of mosses, but a different life-cycle in which the sporophyte is more developed than the gametophyte. Genera such as *Rhynia* have a similar life-cycle but have simple tracheids and so are a kind of vascular plant.

During the Devonian period, vascular plants diversified and spread to many different land environments. In addition to vascular tissues which transport water throughout the body, tracheophytes have an outer layer or cuticle that resists drying out. The sporophyte is the dominant generation, and in modern species develops leaves, stems and roots, while the gametophyte remains very small.

Lycophytes and Euphyllophytes

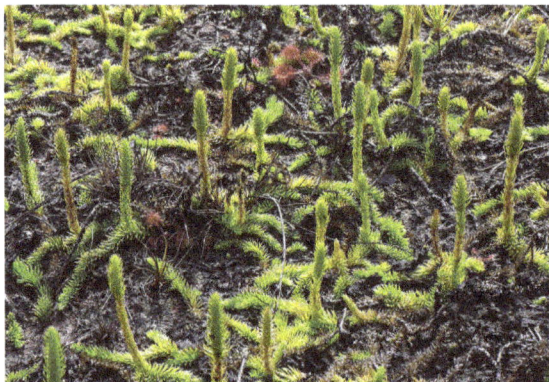

Lycopodiella inundata, a lycophyte

All the vascular plants which disperse through spores were once thought to be related (and were often grouped as 'ferns and allies'). However, recent research suggests that leaves evolved quite separately in two different lineages. The lycophytes or lycopodiophytes – modern clubmosses, spikemosses and quillworts – make up less than 1% of living vascular plants. They have small leaves, often called 'microphylls' or 'lycophylls', which are borne all along the stems in the club-mosses and spikemosses, and which effectively grow from the base, via an intercalary meristem. It is believed that microphylls evolved from outgrowths on stems, such as spines, which later acquired veins (vascular traces).

Although the living lycophytes are all relatively small and inconspicuous plants, more common in the moist tropics than in temperate regions, during the Carboniferous period tree-like lycophytes (such as *Lepidodendron*) formed huge forests that dominated the landscape.

The euphyllophytes, making up more than 99% of living vascular plant species, have large 'true' leaves (megaphylls), which effectively grow from the sides or the apex, via marginal or apical meristems. One theory is that megaphylls developed from three-dimensional branching systems by first 'planation' – flattening to produce a two dimensional branched structure – and then 'webbing' – tissue growing out between the flattened branches. Others have questioned whether megaphylls developed in the same way in different groups.

Ferns and Horsetails

Athyrium filix-femina, unrolling young frond

Euphyllophytes are divided into two lineages: the ferns and horsetails (monilophytes) and the seed plants (spermatophytes). Like all the preceding groups, the monilophytes continue to use spores as their main method of dispersal. Traditionally, whisk ferns and horsetails were treated as distinct from 'true' ferns. Recent research suggests that they all belong together, although there are differences of opinion on the exact classification to be used. Living whisk ferns and horsetails do not have the large leaves (megaphylls) which would be expected of euphyllophytes. However, this has probably resulted from reduction, as evidenced by early fossil horsetails, in which the leaves are broad with branching veins.

Ferns are a large and diverse group, with some 12,000 species. A stereotypical fern has broad, much divided leaves, which grow by unrolling.

Seed Plants

Conifer forest in Northern California

Large seed of a horse chestnut, *Aesculus hippocastanum*

Seed plants, which first appeared in the fossil record towards the end of the Paleozoic era, reproduce using desiccation-resistant capsules called seeds. Starting from a plant which disperses by spores, highly complex changes are needed to produce seeds. The sporophyte has two kinds of spore-forming organs (sporangia). One kind, the megasporangium, produces only a single large spore (a megaspore). This sporangium is surrounded by one or more sheathing layers (integuments) which form the seed coat. Within the seed coat, the megaspore develops into a tiny gametophyte, which in turn produces one or more egg cells. Before fertilization, the sporangium and its contents plus its coat is called an 'ovule'; after fertilization a 'seed'. In parallel to these developments, the other kind of sporangium, the microsporangium, produces microspores. A tiny gametophyte develops inside the wall of a microspore, producing a pollen grain. Pollen grains can be physically transferred between plants by the wind or animals, most commonly insects. Pollen grains can also transfer to an ovule of the same plant, either with the same flower or between two flowers of the same plant (self-fertilization). When a pollen grain reaches an ovule, it enters via a microscopic gap in the coat (the micropyle). The tiny gametophyte inside the pollen grain then produces sperm cells which move to the egg cell and fertilize it. Seed plants include two groups with living members, the gymnosperms and the angiosperms or flowering plants. In gymnosperms, the ovules or seeds are not further enclosed. In angiosperms, they are enclosed in ovaries. A split ovary with a visible seed can be seen in the adjacent image. Angiosperms typically also have other, secondary structures, such as petals, which together form a flower.

Extant seed plants are divided into five groups:

Gymnosperms

- Pinophyta - conifers

- Cycadophyta - cycads

- Ginkgophyta - ginkgo

- Gnetophyta - gnetophytes

Angiosperms

- Magnoliophyta – flowering plants

Bryophyte

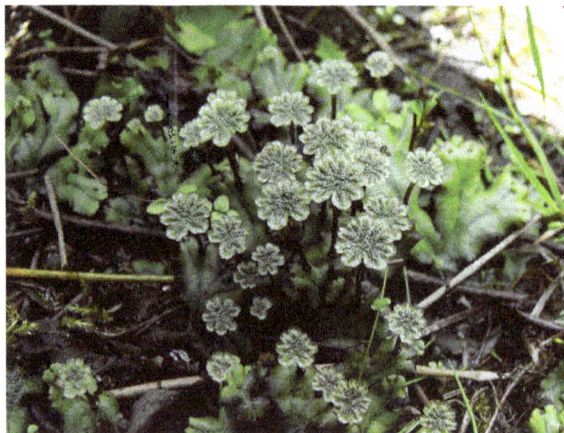

Marchantia, an example of a liverwort.

Bryophytes are an informal group consisting of three divisions of non-vascular land plants (embryophytes), the liverworts, hornworts and mosses. They are characteristically limited in size and prefer moist habitats although they can survive in drier environments. The bryophytes consist of about 20,000 plant species. Bryophytes produce enclosed reproductive structures (gametangia and sporangia), but they do not produce flowers or seeds. They reproduce via spores. Bryophytes are usually considered to be a paraphyletic group and not a monophyletic group, although some studies have produced contrary results. Regardless of their status, the name is convenient and remains in use as an informal collective term. The term "bryophyte" comes from Greek *bryon* "tree-moss, oyster-green" + *phyton* "plant".

The defining features of bryophytes are:

- Their life cycles are dominated by the gametophyte stage

- Their sporophytes are unbranched

- They do not have a true vascular tissue containing lignin (although some have specialized tissues for the transport of water)

Habitat

Bryophytes exist in a wide variety of habitats. They can be found growing in a range of temperatures (cold arctics and in hot deserts), elevations (sea-level to alpine), and moisture (dry deserts to wet rainforests).

Bryophytes can grow where vascularized plants cannot because they do not depend on roots for an uptake of nutrients from soil. Bryophytes can survive on rocks and bare soil.

Life Cycle

Like all land plants (embryophytes), bryophytes have life cycles with alternation of generations. In each cycle, a haploid gametophyte, each of whose cells contains a fixed number of unpaired chromosomes, alternates with a diploid sporophyte, whose cell contain two sets of paired chromosomes. Gametophytes produce haploid sperm and eggs which fuse to form diploid zygotes that grow into sporophytes. Sporophytes produce haploid spores by meiosis, that grow into gametophytes.

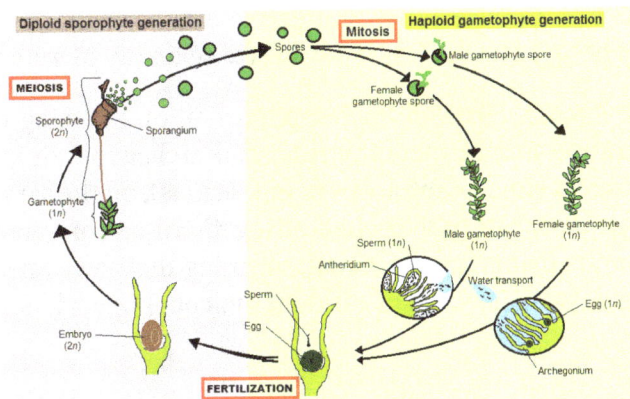

The life cycle of a dioicous bryophyte. The gametophyte (haploid) structures are shown in green, the sporophyte (diploid) in brown.

Bryophytes are gametophyte dominant, meaning that the more prominent, longer-lived plant is the haploid gametophyte. The diploid sporophytes appear only occasionally and remain attached to and nutritionally dependent on the gametophyte. In bryophytes, the sporophytes are always unbranched and produce a single sporangium (spore producing capsule).

Liverworts, mosses and hornworts spend most of their lives as gametophytes. Gametangia (gamete-producing organs), archegonia and antheridia, are produced on the gametophytes, sometimes at the tips of shoots, in the axils of leaves or hidden under thalli. Some bryophytes, such as the liverwort *Marchantia*, create elaborate structures to bear the gametangia that are called gametangiophores. Sperm are flagellated and must swim from the antheridia that produce them to archegonia which may be on a different plant. Arthropods can assist in transfer of sperm.

Fertilized eggs become zygotes, which develop into sporophyte embryos inside the archegonia. Mature sporophytes remain attached to the gametophyte. They consist of a stalk called a seta and a single sporangium or capsule. Inside the sporangium, haploid spores are produced by meiosis. These are dispersed, most commonly by wind, and if they land in a suitable environment can develop into a new gametophyte. Thus bryophytes disperse by a combination of swimming sperm and spores, in a manner similar to lycophytes, ferns and other cryptogams.

Sexuality

The arrangement of antheridia and archegonia on an individual bryophyte plant is usually constant within a species, although in some species it may depend on environmental conditions. The main division is between species in which the antheridia and archegonia occur on the same plant and those in which they occur on different plants. The term monoicous may be used where antheridia and archegonia occur on the same gametophyte and the term dioicous where they occur on different gametophytes.

In seed plants, "monoecious" is used where flowers with anthers (microsporangia) and flowers with ovules (megasporangia) occur on the same sporophyte and "dioecious" where they occur on different sporophytes. These terms occasionally may be used instead of "monoicous" and "dioicous" to describe bryophyte gametophytes. "Monoecious" and "monoicous" are both derived from the Greek for "one house", "dioecious" and "dioicous" from the Greek for two houses. The use of the "oicy" terminology is said to have the advantage of emphasizing the difference between the gametophyte sexuality of bryophytes and the sporophyte sexuality of seed plants.

Monoicous plants are necessarily bisexual (or hermaphroditic), meaning that the same plant has both sexes. The exact arrangement of the antheridia and archegonia in monoicous plants varies. They may be borne on different shoots (autoicous or autoecious), on the same shoot but not together in a common structure (paroicous or paroecious), or together in a common "inflorescence" (synoicous or synoecious). Dioicous plants are unisexual, meaning that the same plant has only one sex. All four patterns (autoicous, paroicous, synoicous and dioicous) occur in species of the moss genus *Bryum*.

Classification and Phylogeny

Traditionally, all living land plants without vascular tissues were classified in a single taxonomic group, often a division (or phylum). More recently, phylogenetic research has questioned whether the bryophytes form a monophyletic group and thus whether they should form a single taxon.

Although a 2005 study supported the traditional view that the bryophytes form a monophyletic group, by 2010 a broad consensus had emerged among systematists that bryophytes as a whole are not a natural group (i.e., are paraphyletic), although each of the three extant (living) groups is monophyletic.

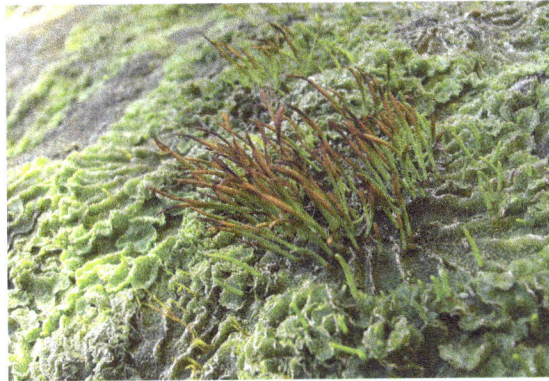

Hornworts include those bryophytes that are believed to be the closest living relatives of the vascular plants.

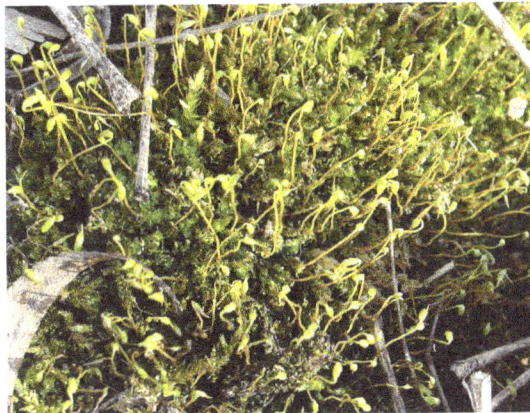

Mosses are one group of bryophytes.

The three bryophyte clades are the Divisions Marchantiophyta (liverworts), Bryophyta (mosses) and Anthocerotophyta (hornworts). The vascular plants or tracheophytes form a fourth, unranked clade of land plants called the "Polysporangiophyta". In this analysis, hornworts are sister to vascular plants and liverworts are sister to all other land plants, including the hornworts and mosses. Phylogenetic studies continue to produce conflicting results. In particular those based on gene sequences suggest the bryophytes are paraphyletic, whereas those based on the amino acid translations of the same genes suggest they are monophyletic. A 2014 study concluded that composition biases were responsible for these differences and that the bryophytes are monophyletic. The issue remains unresolved.

Paraphyletic View

When extinct plants are taken into account, the picture is slightly altered. Some extinct land plants, such as the horneophytes, are not bryophytes, but also are not vascular plants because, like bryophytes, they do not have true vascular tissue. A different distinction is needed. In bryophytes, the sporophyte is a simple unbranched structure with a single spore-forming organ (sporangium). In all other land plants, the polysporangiophytes, the sporophyte is branched and carries many sporangia. It has been argued that this contrast between bryophytes and other land plants is less misleading than the traditional one of non-vascular versus vascular plant, since many mosses have well-developed water-conducting vessels.

The term "bryophyte" thus refers to a grade of lineages defined primarily by what they lack. Compared to other living land plants, they lack vascular tissue containing lignin and branched sporophytes bearing multiple sporangia. The prominence of the gametophyte in the life cycle is also a shared feature of the three bryophyte lineages (extant vascular plants are all sporophyte dominant).

Other Views

An alternative phylogeny, based on amino acids rather than genes.

If this phylogeny proves correct, then the complex sporophyte of living vascular plants might have evolved independently of the simpler unbranched sporophyte present in bryophytes. Other studies suggest a monophyletic group comprising liverworts and mosses, with hornworts being sister to vascular plants.

Evolution

Between 510 - 630 million years ago, land plants evolved from aquatic plants, specifically green algae. Molecular phylogenetic studies conclude that bryophytes are the earliest diverging lineages of the extant land plants. They provide insights into the migration of plants from aquatic environments to land. A number of physical features link bryophytes to both land plants and aquatic plants.

Similarities to Land Plants

Distinct adaptations observed in bryophytes have allowed plants to colonize Earth's terrestrial environments. To prevent desiccation of plant tissues in a terrestrial environment, a waxy cuticle covering the soft tissue of the plant provides protection. The development of gametangia provided further protection specifically for gametes. They also have embryonic development which is a significant adaptation seen in land plants and not green algae. While bryophytes have no truly vascularized tissue, they do have organs that have specific functions, similar to those functions of leaves and stems in higher level land plants.

Similarities to Aquatic Plants

Bryophytes also exhibit connections to their aquatic ancestry. They share various features with their green algae ancestors. Both green algae and bryophytes have chlorophyll a and b, and the chloroplast structures are similar. Like algae and land plants, bryophytes also produce starch and contain cellulose in their walls.

Bryophytes depend on water for reproduction and survival. A thin layer of water is required on the surface of the plant to enable the movement of sperm between gametophytes and the fertilization of an egg.

Comparative Morphology

Summary of the morphological characteristics of the gametophytes of the three groups of bryophytes:

	Liverworts	Mosses	Hornworts
Structure	Thalloid or Foliose	Foliose	Thalloid
Symmetry	Dorsiventral or radial	Radial	Dorsiventral
Rhizoids	Unicellular	Pluricellular	Unicellular
Chloroplasts/cell	Many	Many	One
Protonemata	Reduced	Present	Absent
Gametangia (antheridia and archegonia)	Superficial	Superficial	Immersed

Summary of the morphological characteristics of the sporophytes of the three groups of bryophytes:

	Liverworts	Mosses	Hornworts
Stomata	Absent	Present	Present
Structure	Small, without chlorophyll	Large, with chlorophyll	Large, with chlorophyll
Persistence	Ephemeral	Persistent	Persistent
Growth	Defined	Defined	Continuous
Seta	Present	Present	Absent
Capsule form	Simple	Differentiated (operculum, peristome)	Elongated
Maturation of spores	Simultaneous	Simultaneous	Graduate
Dispersion of spores	Elaters	Peristome teeth	Pseudo-elaters
Columella	Absent	Present	Present
Dehiscence	Longitudinal or irregular	Transverse	Longitudinal

Uses

Environmental

- Soil Conditioning
- Bioindicators
- Moss gardens
- Pesticides

Characteristics of bryophytes make them useful to the environment. Depending on the specific plant texture, bryophytes have been shown to help improve the water retention and air space within soil. Bryophytes are used in pollution studies to indicate soil pollution (such as the presence of heavy metals), air pollution, and UV-B radiation. Gardens in Japan are designed with moss to create peacful sanctuaries. Some bryophytes have been found to produce natural pesticides. The liverwort, *Plagiochila,* produces a chemical that is poisonous to mice. Other bryophytes produce chemicals that are antifeedants which protect them from being eaten by slugs. When *Phythium sphagnum* is sprinkled on the soil of germinating seeds, it inhibits growth of "damping off fungus" which would otherwise kill young seedlings.

Moss peat is made from *Sphagnum*

Commercial

- Fuel

- Packaging

- Wound Dressing

Peat is a fuel that is produced from dried bryophytes, typically *sphagnum.* Bryophytes antibiotic properties and ability to retain water make them a useful packaging material for vegetables, flowers, and bulbs. Also, because of the antibiotic properties, *sphagnum* was used as a surgical dressing in World War I.

Gymnosperm

The gymnosperms are a group of seed-producing plants (spermatophytes) that includes conifers (Pinophyta), cycads, *Ginkgo*, and gnetophytes. The term "gymnosperm" comes from the Greek composite word (gymnos, naked and sperma, seed), meaning naked seeds. The name is based on the unenclosed condition of their seeds (called ovules in their unfertilized state). The non-encased condition of their seeds stands in contrast to the seeds and ovules of flowering plants (angiosperms), which are enclosed within an ovary. Gymnosperm seeds develop either on the surface of scales or leaves, which are often modified to form cones, or solitary as in Yew, *Torreya, Ginkgo.*

Encephalartos sclavoi cone, about 30 cm long

The gymnosperms and angiosperms together compose the spermatophytes or seed plants. The gymnosperms are divided into six phyla. Organisms that belong to the Cycadophyta, Ginkgophyta, Gnetophyta, and Pinophyta (also known as Coniferophyta) phyla are still in existence while those in the Pteridospermales and Cordaitales phyla are now extinct.

By far the largest group of living gymnosperms are the conifers (pines, cypresses, and relatives), followed by cycads, gnetophytes (*Gnetum*, *Ephedra* and *Welwitschia*), and *Ginkgo biloba* (a single living species).

Classification

In early classification schemes, the gymnosperms (Gymnospermae) were regarded as a "natural" group. There is conflicting evidence on the question of whether the living gymnosperms form a clade. The fossil record of gymnosperms includes many distinctive taxa that do not belong to the four modern groups, including seed-bearing trees that have a somewhat fern-like vegetative morphology (the so-called "seed ferns" or pteridosperms.) When fossil gymnosperms such as Bennettitales, *Caytonia* and the glossopterids are considered, it is clear that angiosperms are nested within a larger gymnosperm clade, although which group of gymnosperms is their closest relative remains unclear.

There are 12 families, 83 known genera with a total of ca 1080 known species (Christenhusz & Byng 2016).

Subclass Cycadidae

- Order Cycadales

  o   Family Cycadaceae: *Cycas*

- o Family Zamiaceae: *Dioon, Bowenia, Macrozamia, Lepidozamia, Encephalartos, Stangeria, Ceratozamia, Microcycas, Zamia.*

Subclass Ginkgoidae

- Order Ginkgoales

  - o Family Ginkgoaceae: *Ginkgo*

Subclass Gnetidae

- Order Welwitschiales

  - o Family Welwitschiaceae: *Welwitschia*

- Order Gnetales

  - o Family Gnetaceae: *Gnetum*

- Order Ephedrales

  - o Family Ephedraceae: *Ephedra*

Subclass Pinidae

- Order Pinales

  - o Family Pinaceae: *Cedrus, Pinus, Cathaya, Picea, Pseudotsuga, Larix, Pseudolarix, Tsuga, Nothotsuga, Keteleeria, Abies*

- Order Araucariales

  - o Family Araucariaceae: *Araucaria, Wollemia, Agathis*

  - o Family Podocarpaceae: *Phyllocladus, Lepidothamnus, Prumnopitys, Sundacarpus, Halocarpus, Parasitaxus, Lagarostrobos, Manoao, Saxegothaea, Microcachrys, Pherosphaera, Acmopyle, Dacrycarpus, Dacrydium, Falcatifolium, Retrophyllum, Nageia, Afrocarpus, Podocarpus*

- Order Cupressales

  - o Family Sciadopityaceae: *Sciadopitys*

  - o Family Cupressaceae: *Cunninghamia, Taiwania, Athrotaxis, Metasequoia, Sequoia, Sequoiadendron, Cryptomeria, Glyptostrobus, Taxodium, Papuacedrus, Austrocedrus, Libocedrus, Pilgerodendron, Widdringtonia, Diselma, Fitzroya, Callitris* (incl. *Actinostrobus* and *Neocallitropsis*)*, Thujopsis, Thuja, Fokienia, Chamaecyparis, Callitropsis, Cupressus, Juniperus, Xanthocyparis, Calocedrus, Tetraclinis, Platycladus, Microbiota*

  - o Family Taxaceae: *Austrotaxus, Pseudotaxus, Taxus, Cephalotaxus, Amentotaxus, Torreya*

Diversity and Origin

There are more than 1000 extant or currently living species of gymnosperms in 88 plant genera belonging to 14 plant families.

It is widely accepted that the gymnosperms originated in the late Carboniferous period, replacing the lycopsid rainforests of the tropical region. This appears to have been the result of a whole genome duplication event around 319 million years ago. Early characteristics of seed plants were evident in fossil progymnosperms of the late Devonian period around 383 million years ago. It has been suggested that during the mid-Mesozoic era, pollination of some extinct groups of gymnosperms was by extinct species of scorpionflies that had specialized proboscis for feeding on pollination drops. The scorpionflies likely engaged in pollination mutualisms with gymnosperms, long before the similar and independent coevolution of nectar-feeding insects on angiosperms. Evidence has also been found that mid-Mesozoic gymnosperms were pollinated by Kalligrammatid lacewings, a now-extinct genus with members which (in an example of convergent evolution) resembled the modern butterflies that arose far later.

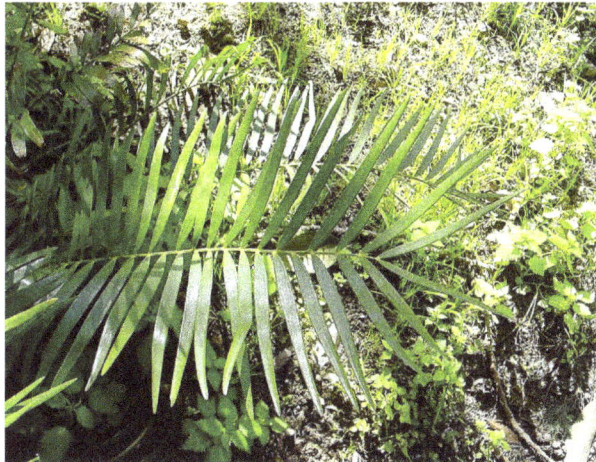

Zamia integrifolia, a cycad native to Florida

Conifers are by far the most abundant extant group of gymnosperms with six to eight families, with a total of 65-70 genera and 600-630 species (696 accepted names). Conifers are woody plants and most are evergreens. The leaves of many conifers are long, thin and needle-like, other species, including most Cupressaceae and some Podocarpaceae, have flat, triangular scale-like leaves. *Agathis* in Araucariaceae and *Nageia* in Podocarpaceae have broad, flat strap-shaped leaves.

Cycads are the next most abundant group of gymnosperms, with two or three families, 11 genera, and approximately 338 species. A majority of cycads are native to tropical climates and are most abundantly found in regions near the equator. The other extant groups are the 95-100 species of Gnetales and one species of Ginkgo.

Uses

Gymnosperms have major economic uses. Pine, fir, spruce, and cedar are all examples of conifers that are used for lumber, paper production, and resin. Some other common uses for gymnosperms are soap, varnish, nail polish, food, gum, and perfumes.

Life Cycle

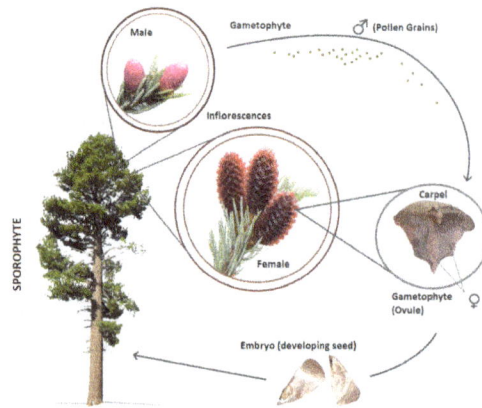

Example of gymnosperm lifecycle

Gymnosperms, like all vascular plants, have a sporophyte-dominant life cycle, which means they spend most of their life cycle with diploid cells, while the gametophyte (gamete-bearing phase) is relatively short-lived. Two spore types, microspores and megaspores, are typically produced in pollen cones or ovulate cones, respectively. Gametophytes, as with all heterosporous plants, develop within the spore wall. Pollen grains (microgametophytes) mature from microspores, and ultimately produce sperm cells. Megagametophytes develop from megaspores and are retained within the ovule. Gymnosperms produce multiple archegonia, which produce the female gamete. During pollination, pollen grains are physically transferred between plants from the pollen cone to the ovule. Pollen is usually moved by wind or insects. Whole grains enter each ovule through a microscopic gap in the ovule coat (integument) called the micropyle. The pollen grains mature further inside the ovule and produce sperm cells. Two main modes of fertilization are found in gymnosperms. Cycads and *Ginkgo* have motile sperm that swim directly to the egg inside the ovule, whereas conifers and gnetophytes have sperm with no flagella that are moved along a pollen tube to the egg. After syngamy (joining of the sperm and egg cell), the zygote develops into an embryo (young sporophyte). More than one embryo is usually initiated in each gymnosperm seed. The mature seed comprises the embryo and the remains of the female gametophyte, which serves as a food supply, and the seed coat.

Genetics

The first published sequenced genome for any gymnosperm was the genome of *Picea abies* in 2013.

Cycad

Cycads are seed plants with a long fossil history that were formerly more abundant and more diverse than they are today. They typically have a stout and woody (ligneous) trunk with a crown of large, hard and stiff, evergreen leaves. They usually have pinnate leaves. The individual plants are either all male or all female (dioecious). Cycads vary in size from having trunks only a few centimeters to several meters tall. They typically grow very slowly and live very long, with some specimens known to be as much as 1,000 years old. Because of their superficial resemblance, they are sometimes mistaken for palms or ferns, but are only distantly related to either group.

Cycads are gymnosperms (naked seeded), meaning their unfertilized seeds are open to the air to be directly fertilized by pollination, as contrasted with angiosperms, which have enclosed seeds with more complex fertilization arrangements. Cycads have very specialized pollinators, usually a specific species of beetle. They have been reported to fix nitrogen in association with various cyanobacteria living in the roots (the "coralloid" roots). These photosynthetic bacteria produce a neurotoxin called BMAA that is found in the seeds of cycads. This neurotoxin may enter a human food chain as the cycad seeds may be eaten directly as a source of flour by humans or by wild or feral animals such as bats, and humans may eat these animals. It is hypothesized that this is a source of some neurological diseases in humans.

Description

Cycads have a rosette of pinnate leaves around cylindrical trunk

Cycads have a cylindrical trunk which usually does not branch. Leaves grow directly from the trunk, and typically fall when older, leaving a crown of leaves at the top. The leaves grow in a rosette form, with new foliage emerging from the top and center of the crown. The trunk may be buried, so the leaves appear to be emerging from the ground, so the plant appears to be a basal rosette. The leaves are generally large in proportion to the trunk size, and sometimes even larger than the trunk.

The leaves are pinnate (in the form of bird feathers, pinnae), with a central leaf stalk from which parallel "ribs" emerge from each side of the stalk, perpendicular to it. The leaves are typically either compound (the leaf stalk has leaflets emerging from it as "ribs"), or have edges (margins) so deeply cut (incised) so as to appear compound. Some species have leaves that are bipinnate, which means the leaflets each have their own subleaflets, growing in the same form on the leaflet as the leaflets grow on the stalk of the leaf (self-similar geometry).

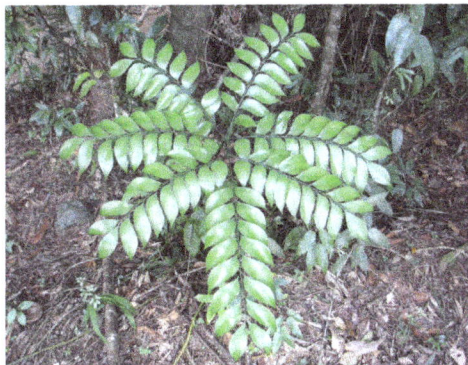

Bowenia spectabilis : plant with single frond in the Daintree rainforest,
north-east Queensland

Taxonomy

The three extant families of cycads all belong to the order Cycadales, and are Cycadaceae, Stangeriaceae, and Zamiaceae. These cycads have changed little since the Jurassic, compared to some major evolutionary changes in other plant divisions. Five additional families belonging to the Medullosales became extinct by the end of the Paleozoic Era.

Cycads are most closely related to the extinct Bennettitales, and are also relatively close relatives to the Ginkgoales, as shown in the following phylogeny:

Classification of the Cycadophyta to the rank of family.

Class Cycadopsida

      Order Medullosales †

            Family Alethopteridaceae

            Family Cyclopteridaceae

            Family Neurodontopteridaceae

            Family Parispermaceae

      Order Cycadales

            Suborder Cycadineae

                  Family Cycadaceae

            Suborder Zamiineae

                  Family Stangeriaceae

                  Family Zamiaceae

Historical Diversity

The probable former range of cycads can be inferred from their global distribution. For example, the family Stangeriaceae only contains three extant species in Africa and Australia. Diverse fossils of this family have been dated to 135 mya, indicating that diversity may have been much greater before the Jurassic and late Triassic mass extinction events. However, the cycad fossil record is generally poor and little can be deduced about the effects of each mass extinction event on their diversity.

Instead, correlations can be made between the number of extant gymnosperms and angiosperms. It is likely that cycad diversity was affected more by the great angiosperm radiation in the mid-Cretaceous than by extinctions. Very slow cambial growth was first used to define cycads, and because of this characteristic the group could not compete with the rapidly growing, relatively short-lived angiosperms, which now number over 250,000 species, compared to the 947 remaining gymnosperms. It is surprising that the cycads are still extant, having been faced with extreme competition and five major extinctions. The ability of cycads to survive in relatively dry environments where plant diversity is generally lower, may explain their long persistence and longevity.

Origins

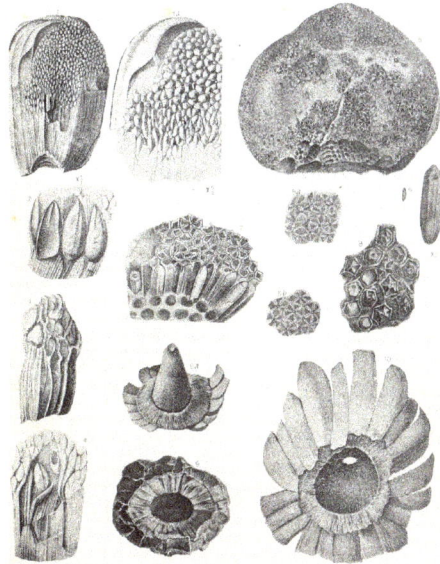

Fossil images of Cycads found in America (1906)

The cycad fossil record dates to the early Permian, 280 million years ago (mya). There is controversy over older cycad fossils that date to the late Carboniferous period, 300–325 mya. This clade probably diversified extensively within its first few million years, although the extent to which it radiated is unknown because relatively few fossil specimens have been found. The regions to which cycads are restricted probably indicate their former distribution in the Pangea before the supercontinents Laurasia and Gondwana separated. Recent studies have indicated the common perception of existing cycad species as living fossils is largely misplaced, with only *Bowenia* dating to the Cretaceous or earlier. Although the cycad lineage itself is ancient, most extant species have evolved in the last 12 million years.

The family Stangeriaceae (named for Dr. William Stanger, 1811–1854), consisting of only three extant species, is thought to be of Gondwanan origin, as fossils have been found in Lower Cretaceous deposits in Argentina, dating to 70–135 mya. The family Zamiaceae is more diverse, with a fossil record extending from the middle Triassic to the Eocene (54–200 mya) in North and South America, Europe, Australia, and Antarctica, implying the family was present before the break-up of Pangea. The family Cycadaceae is thought to be an early offshoot from other cycads, with fossils from Eocene deposits (38–54 mya) in Japan, China, and North America, indicating this family originated in Laurasia. *Cycas* is the only genus in the family and contains 99 species, the most of any cycad genus. Molecular data have recently shown *Cycas* species in Australasia and the east coast of Africa are recent arrivals, suggesting adaptive radiation may have occurred. The current distribution of cycads may be due to radiations from a few ancestral types sequestered on Laurasia and Gondwana, or could be explained by genetic drift following the separation of already evolved genera. Both explanations account for the strict endemism across present continental lines.

Distribution

The living cycads are found across much of the subtropical and tropical parts of the world. The greatest diversity occurs in South and Central America. They are also found in Mexico, the Antilles, southeastern United States, Australia, Melanesia, Micronesia, Japan, China, Southeast Asia, India, Sri Lanka, Madagascar, and southern and tropical Africa, where at least 65 species occur. Some can survive in harsh desert or semi-desert climates (xerophytic), others in wet rain forest conditions, and some in both. Some can grow in sand or even on rock, some in oxygen-poor, swampy, bog-like soils rich in organic material. Some are able to grow in full sun, some in full shade, and some in both. Some are salt tolerant (halophytes).

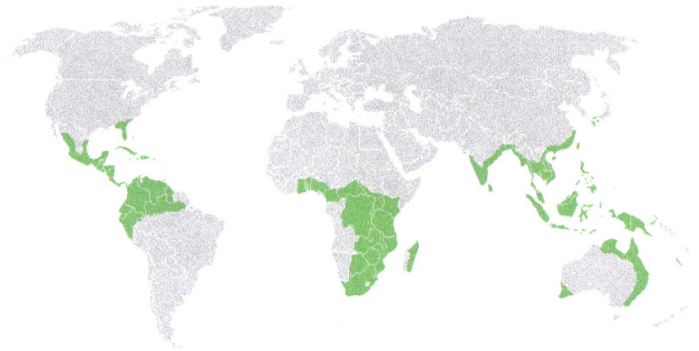

Approximate world distribution of living Cycadales

Species diversity of the extant cycads peaks at 17° 15"N and 28° 12"S, with a minor peak at the equator. There is therefore not a latitudinal diversity gradient towards the equator but towards the Tropic of Cancer and the Tropic of Capricorn. However, the peak near the northern tropic is largely due to *Cycas* in Asia and *Zamia* in the New World, whereas the peak near the southern tropic is due to *Cycas* again, and also to the diverse genus *Encephalartos* in southern and central Africa, and *Macrozamia* in Australia. Thus, the distribution pattern of cycad species with latitude appears to be an artifact of the geographical isolation of the remaining cycad genera and their species, and perhaps because they are partly xerophytic rather than simply tropical.

Flowering Plant

The flowering plants (angiosperms), also known as Angiospermae or Magnoliophyta, are the most diverse group of land plants, with 416 families, approx. 13,164 known genera and a total of c. 295,383 known species. Like gymnosperms, angiosperms are seed-producing plants; they are distinguished from gymnosperms by characteristics including flowers, endosperm within the seeds, and the production of fruits that contain the seeds. Etymologically, angiosperm means a plant that produces seeds within an enclosure, in other words, a fruiting plant. The term "angiosperm" comes from the Greek composite word (*angeion*, "case" or "casing", and *sperma*, "seed") meaning "enclosed seeds", after the enclosed condition of the seeds.

The ancestors of flowering plants diverged from gymnosperms in the Triassic Period, during the range 245 to 202 million years ago (mya), and the first flowering plants are known from 160 mya.

They diversified extensively during the Lower Cretaceous, became widespread by 120 mya, and replaced conifers as the dominant trees from 100 to 60 mya.

Description

Angiosperm Derived Characteristics

Bud of a pink rose

Angiosperms differ from other seed plants in several ways, described in the table. These distinguishing characteristics taken together have made the angiosperms the most diverse and numerous land plants and the most commercially important group to humans.

Distinctive features of Angiosperms	
Feature	**Description**
Flowering organs	Flowers, the reproductive organs of flowering plants, are the most remarkable feature distinguishing them from the other seed plants. Flowers provided angiosperms with the means to have a more species-specific breeding system, and hence a way to evolve more readily into different species without the risk of crossing back with related species. Faster speciation enabled the Angiosperms to adapt to a wider range of ecological niches. This has allowed flowering plants to largely dominate terrestrial ecosystems.
Stamens with two pairs of pollen sacs	Stamens are much lighter than the corresponding organs of gymnosperms and have contributed to the diversification of angiosperms through time with adaptations to specialized pollination syndromes, such as particular pollinators. Stamens have also become modified through time to prevent self-fertilization, which has permitted further diversification, allowing angiosperms eventually to fill more niches.
Reduced male parts, three cells	The male gametophyte in angiosperms is significantly reduced in size compared to those of gymnosperm seed plants. The smaller size of the pollen reduces the amount of time between pollination — the pollen grain reaching the female plant — and fertilization. In gymnosperms, fertilization can occur up to a year after pollination, whereas in angiosperms, fertilization begins very soon after pollination. The shorter amount of time between pollination and fertilization allows angiosperms to produce seeds earlier after pollination than gymnosperms, providing angiosperms a distinct evolutionary advantage.

Closed carpel enclosing the ovules (carpel or carpels and accessory parts may become the fruit)	The closed carpel of angiosperms also allows adaptations to specialized pollination syndromes and controls. This helps to prevent self-fertilization, thereby maintaining increased diversity. Once the ovary is fertilized, the carpel and some surrounding tissues develop into a fruit. This fruit often serves as an attractant to seed-dispersing animals. The resulting cooperative relationship presents another advantage to angiosperms in the process of dispersal.
Reduced female gametophyte, seven cells with eight nuclei	The reduced female gametophyte, like the reduced male gametophyte, may be an adaptation allowing for more rapid seed set, eventually leading to such flowering plant adaptations as annual herbaceous life-cycles, allowing the flowering plants to fill even more niches.
Endosperm	In general, endosperm formation begins after fertilization and before the first division of the zygote. Endosperm is a highly nutritive tissue that can provide food for the developing embryo, the cotyledons, and sometimes the seedling when it first appears.

Vascular Anatomy

The amount and complexity of tissue-formation in flowering plants exceeds that of gymnosperms. The vascular bundles of the stem are arranged such that the xylem and phloem form concentric rings.

In the dicotyledons, the bundles in the very young stem are arranged in an open ring, separating a central pith from an outer cortex. In each bundle, separating the xylem and phloem, is a layer of meristem or active formative tissue known as cambium. By the formation of a layer of cambium between the bundles (interfascicular cambium), a complete ring is formed, and a regular periodical increase in thickness results from the development of xylem on the inside and phloem on the outside. The soft phloem becomes crushed, but the hard wood persists and forms the bulk of the stem and branches of the woody perennial. Owing to differences in the character of the elements produced at the beginning and end of the season, the wood is marked out in transverse section into concentric rings, one for each season of growth, called annual rings.

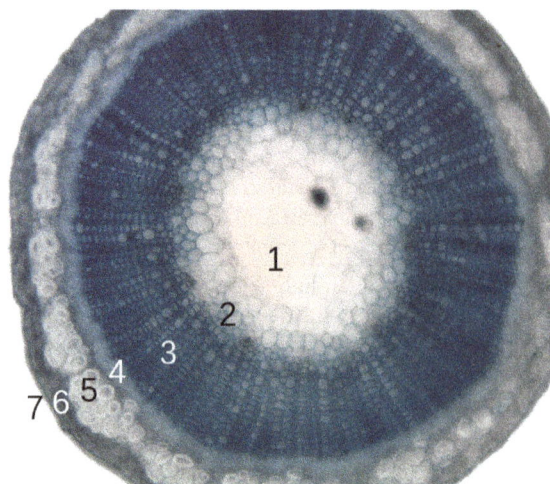

Cross-section of a stem of the angiosperm flax:
1. Pith, 2. Protoxylem, 3. Xylem I, 4. Phloem I, 5. Sclerenchyma (bast fibre), 6. Cortex, 7. Epidermis

Among the monocotyledons, the bundles are more numerous in the young stem and are scattered through the ground tissue. They contain no cambium and once formed the stem increases in diameter only in exceptional cases.

Reproductive Anatomy

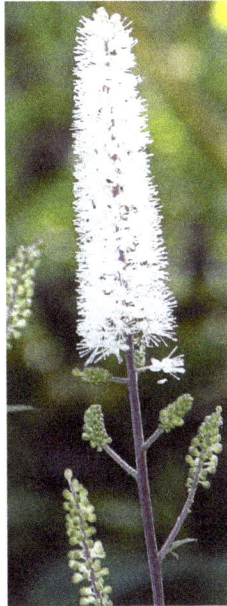

A collection of flowers forming an inflorescence

The characteristic feature of angiosperms is the flower. Flowers show remarkable variation in form and elaboration, and provide the most trustworthy external characteristics for establishing relationships among angiosperm species. The function of the flower is to ensure fertilization of the ovule and development of fruit containing seeds. The floral apparatus may arise terminally on a shoot or from the axil of a leaf (where the petiole attaches to the stem). Occasionally, as in violets, a flower arises singly in the axil of an ordinary foliage-leaf. More typically, the flower-bearing portion of the plant is sharply distinguished from the foliage-bearing or vegetative portion, and forms a more or less elaborate branch-system called an inflorescence.

There are two kinds of reproductive cells produced by flowers. Microspores, which will divide to become pollen grains, are the "male" cells and are borne in the stamens (or microsporophylls). The "female" cells called megaspores, which will divide to become the egg cell (megagametogenesis), are contained in the ovule and enclosed in the carpel (or megasporophyll).

The flower may consist only of these parts, as in willow, where each flower comprises only a few stamens or two carpels. Usually, other structures are present and serve to protect the sporophylls and to form an envelope attractive to pollinators. The individual members of these surrounding structures are known as sepals and petals (or tepals in flowers such as *Magnolia* where sepals and petals are not distinguishable from each other). The outer series (calyx of sepals) is usually green and leaf-like, and functions to protect the rest of the flower, especially the bud. The inner series (corolla of petals) is, in general, white or brightly colored, and is more delicate in structure. It functions to attract insect or bird pollinators. Attraction is effected by color, scent, and nectar, which may be secreted in some part of the flower. The characteristics that attract pollinators account for the popularity of flowers and flowering plants among humans.

While the majority of flowers are perfect or hermaphrodite (having both pollen and ovule producing parts in the same flower structure), flowering plants have developed numerous morphological and physiological mechanisms to reduce or prevent self-fertilization. Heteromorphic flowers have short carpels and long stamens, or vice versa, so animal pollinators cannot easily transfer pollen to the pistil (receptive part of the carpel). Homomorphic flowers may employ a biochemical (physiological) mechanism called self-incompatibility to discriminate between self and non-self pollen grains. In other species, the male and female parts are morphologically separated, developing on different flowers.

Taxonomy

History of Classification

From 1736, an illustration of Linnaean classification

The botanical term "Angiosperm", from the Ancient Greek, *angeíon* (bottle, vessel) and σπέρμα, (seed), was coined in the form Angiospermae by Paul Hermann in 1690, as the name of one of his primary divisions of the plant kingdom. This included flowering plants possessing seeds enclosed in capsules, distinguished from his Gymnospermae, or flowering plants with achenial or schizo-carpic fruits, the whole fruit or each of its pieces being here regarded as a seed and naked. The term and its antonym were maintained by Carl Linnaeus with the same sense, but with restricted application, in the names of the orders of his class Didynamia. Its use with any approach to its modern scope became possible only after 1827, when Robert Brown established the existence of truly naked ovules in the Cycadeae and Coniferae, and applied to them the name Gymnosperms. From that time onward, as long as these Gymnosperms were, as was usual, reckoned as dicotyledonous flowering plants, the term Angiosperm was used antithetically by botanical writers, with varying scope, as a group-name for other dicotyledonous plants.

In 1851, Hofmeister discovered the changes occurring in the embryo-sac of flowering plants, and determined the correct relationships of these to the Cryptogamia. This fixed the position of Gymnosperms as a class distinct from Dicotyledons, and the term Angiosperm then gradually came to be accepted as the suitable designation for the whole of the flowering plants other than Gymnosperms, including the classes of Dicotyledons and Monocotyledons. This is the sense in which the term is used today.

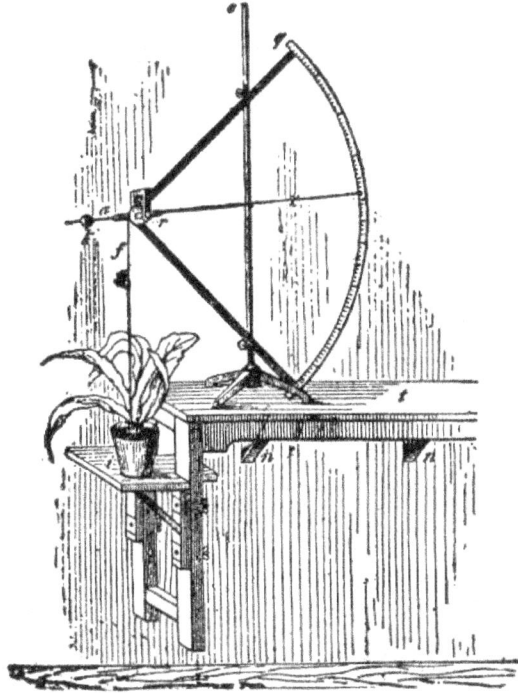

An auxanometer, a device for measuring increase or rate of growth in plants

In most taxonomies, the flowering plants are treated as a coherent group. The most popular descriptive name has been Angiospermae (Angiosperms), with Anthophyta ("flowering plants") a second choice. These names are not linked to any rank. The Wettstein system and the Engler system use the name Angiospermae, at the assigned rank of subdivision. The Reveal system treated flowering plants as subdivision Magnoliophytina (Frohne & U. Jensen ex Reveal, Phytologia 79: 70 1996), but later split it to Magnoliopsida, Liliopsida, and Rosopsida. The Takhtajan system and Cronquist system treat this group at the rank of division, leading to the name Magnoliophyta (from the family name Magnoliaceae). The Dahlgren system and Thorne system (1992) treat this group at the rank of class, leading to the name Magnoliopsida. The APG system of 1998, and the later 2003 and 2009 revisions, treat the flowering plants as a clade called angiosperms without a formal botanical name. However, a formal classification was published alongside the 2009 revision in which the flowering plants form the Subclass Magnoliidae.

The internal classification of this group has undergone considerable revision. The Cronquist system, proposed by Arthur Cronquist in 1968 and published in its full form in 1981, is still widely used but is no longer believed to accurately reflect phylogeny. A consensus about how the flowering plants should be arranged has recently begun to emerge through the work of the Angiosperm Phylogeny Group (APG), which published an influential reclassification of the angiosperms in 1998. Updates incorporating more recent research were published as APG II in 2003 and as APG III in 2009.

Monocot (left) and dicot seedlings

Traditionally, the flowering plants are divided into two groups,

- Dicotyledoneae or Magnoliopsida

- Monocotyledoneae or Liliopsida

which in the Cronquist system are called Magnoliopsida (at the rank of class, formed from the family name Magnoliaceae) and Liliopsida (at the rank of class, formed from the family name Liliaceae). Other descriptive names allowed by Article 16 of the ICBN include Dicotyledones or Dicotyledoneae, and Monocotyledones or Monocotyledoneae, which have a long history of use. In English a member of either group may be called a dicotyledon (plural dicotyledons) and monocotyledon (plural monocotyledons), or abbreviated, as dicot (plural dicots) and monocot (plural monocots). These names derive from the observation that the dicots most often have two cotyledons, or embryonic leaves, within each seed. The monocots usually have only one, but the rule is not absolute either way. From a broad diagnostic point of view, the number of cotyledons is neither a particularly handy nor a reliable character.

Recent studies, as by the APG, show that the monocots form a monophyletic group (clade) but that the dicots do not (they are paraphyletic). Nevertheless, the majority of dicot species do form a monophyletic group, called the eudicots or tricolpates. Of the remaining dicot species, most belong to a third major clade known as the magnoliids, containing about 9,000 species. The rest include a paraphyletic grouping of early branching taxa known collectively as the basal angiosperms, plus the families Ceratophyllaceae and Chloranthaceae.

Modern Classification

There are eight groups of living angiosperms:

- Basal angiosperms (ANA: *Amborella*, Nymphaeales, Austrobaileyales)

  ○ *Amborella*, a single species of shrub from New Caledonia;

  ○ Nymphaeales, about 80 species, water lilies and Hydatellaceae;

  ○ Austrobaileyales, about 100 species of woody plants from various parts of the world

- Core angiosperms (Mesangiospermae)

  ○ Chloranthales, several dozen species of aromatic plants with toothed leaves;

  ○ Magnoliids, about 9,000 species, characterized by trimerous flowers, pollen with one pore, and usually branching-veined leaves—for example magnolias, bay laurel, and black pepper;

  ○ Monocots, about 70,000 species, characterized by trimerous flowers, a single cotyledon, pollen with one pore, and usually parallel-veined leaves—for example grasses, orchids, and palms;

  ○ *Ceratophyllum*, about 6 species of aquatic plants, perhaps most familiar as aquarium plants;

  ○ Eudicots, about 175,000 species, characterized by 4- or 5-merous flowers, pollen with three pores, and usually branching-veined leaves—for example sunflowers, petunia, buttercup, apples, and oaks.

The exact relationship between these eight groups is not yet clear, although there is agreement that the first three groups to diverge from the ancestral angiosperm were Amborellales, Nymphaeales, and Austrobaileyales. The term basal angiosperms refers to these three groups. Among the remaining five groups (core angiosperms), the relationship between the three broadest of these groups (magnoliids, monocots, and eudicots) remains unclear. Zeng and colleagues describe four competing schemes. Of these, eudicots and monocots are the largest and most diversified, with ~ 75% and 20% of angiosperm species, respectively. Some analyses make the magnoliids the first to diverge, others the monocots. *Ceratophyllum* seems to group with the eudicots rather than with the monocots. The 2016 Angiosperm Phylogeny Group revision (APG IV) retained the overall higher order relationship described in APG III.

Evolution

Fossilized spores suggest that higher plants (embryophytes) have lived on land for at least 475 million years. Early land plants reproduced sexually with flagellated, swimming sperm, like the green algae from which they evolved. An adaptation to terrestrialization was the development of upright meiosporangia for dispersal by spores to new habitats. This feature is lacking in the descendants of their nearest algal relatives, the Charophycean green algae. A later terrestrial adaptation took place with retention of the delicate, avascular sexual stage, the gametophyte, within the tissues of the vascular sporophyte. This occurred by spore germination within sporangia rather than spore release, as in non-seed plants. A current example of how this might have happened can be seen in the precocious spore germination in *Selaginella*, the spike-moss. The result for the ancestors of angiosperms was enclosing them in a case, the seed. The first seed bearing plants, like the ginkgo, and conifers (such as pines and firs), did not produce flowers. The pollen grains (male gameto-phytes) of *Ginkgo* and cycads produce a pair of flagellated, mobile sperm cells that "swim" down the developing pollen tube to the female and her eggs.

Flowers of *Malus sylvestris* (crab apple)

Flowers and leaves of *Oxalis pes-caprae* (Bermuda buttercup)

The apparently sudden appearance of nearly modern flowers in the fossil record initially posed such a problem for the theory of evolution that Charles Darwin called it an *"abominable mystery"*. However, the fossil record has considerably grown since the time of Darwin, and recently discovered angiosperm fossils such as *Archaefructus*, along with further discoveries of fossil gymnosperms, suggest how angiosperm characteristics may have been acquired in a series of steps. Several groups of extinct gymnosperms, in particular seed ferns, have been proposed as the ancestors of flowering plants, but there is no continuous fossil evidence showing exactly how flowers evolved. Some older fossils, such as the upper Triassic *Sanmiguelia*, have been suggested. Based on current evidence, some propose that the ancestors of the angiosperms diverged from an unknown group of gymnosperms in the Triassic period (245–202 million years ago). Fossil angiosperm-like pollen from the Middle Triassic (247.2–242.0 Ma) suggests an older date for their origin. A close relationship between angiosperms and gnetophytes, proposed on the basis of morphological evidence, has more recently been disputed on the basis of molecular evidence that suggest gnetophytes are instead more closely related to other gymnosperms.

The evolution of seed plants and later angiosperms appears to be the result of two distinct rounds of whole genome duplication events. These occurred at 319 million years ago and 192 million years ago. Another possible whole genome duplication event at 160 million years ago perhaps created the ancestral line that led to all modern flowering plants. That event was studied by sequencing the genome of an ancient flowering plant, *Amborella trichopoda*, and directly addresses Darwin's *"abominable mystery."*

The earliest known macrofossil confidently identified as an angiosperm, *Archaefructus liaoningensis*, is dated to about 125 million years BP (the Cretaceous period), whereas pollen considered to be of angiosperm origin takes the fossil record back to about 130 million years BP. However, one study has suggested that the early-middle Jurassic plant *Schmeissneria*, traditionally considered a type of ginkgo, may be the earliest known angiosperm, or at least a close relative. In addition, circumstantial chemical evidence has been found for the existence of angiosperms as early as 250 million years ago. Oleanane, a secondary metabolite produced by many flowering plants, has been found in Permian deposits of that age together with fossils of gigantopterids. Gigantopterids are a group of extinct seed plants that share many morphological traits with flowering plants, although they are not known to have been flowering plants themselves.

In 2013 flowers encased in amber were found and dated 100 million years before present. The amber had frozen the act of sexual reproduction in the process of taking place. Microscopic images showed tubes growing out of pollen and penetrating the flower's stigma. The pollen was sticky, suggesting it was carried by insects.

Recent DNA analysis based on molecular systematics showed that *Amborella trichopoda*, found on the Pacific island of New Caledonia, belongs to a sister group of the other flowering plants, and morphological studies suggest that it has features that may have been characteristic of the earliest flowering plants.

The orders Amborellales, Nymphaeales, and Austrobaileyales diverged as separate lineages from the remaining angiosperm clade at a very early stage in flowering plant evolution.

The great angiosperm radiation, when a great diversity of angiosperms appears in the fossil record, occurred in the mid-Cretaceous (approximately 100 million years ago). However, a study in 2007 estimated that the division of the five most recent (the genus *Ceratophyllum*, the family Chloranthaceae, the eudicots, the magnoliids, and the monocots) of the eight main groups occurred around 140 million years ago. By the late Cretaceous, angiosperms appear to have dominated environments formerly occupied by ferns and cycadophytes, but large canopy-forming trees replaced conifers as the dominant trees only close to the end of the Cretaceous 66 million years ago or even later, at the beginning of the Tertiary. The radiation of herbaceous angiosperms occurred much later. Yet, many fossil plants recognizable as belonging to modern families (including beech, oak, maple, and magnolia) had already appeared by the late Cretaceous.

Two bees on a flower head of Creeping Thistle, *Cirsium arvense*

It is generally assumed that the function of flowers, from the start, was to involve mobile animals in their reproduction processes. That is, pollen can be scattered even if the flower is not brightly colored or oddly shaped in a way that attracts animals; however, by expending the energy required to create such traits, angiosperms can enlist the aid of animals and, thus, reproduce more efficiently.

Island genetics provides one proposed explanation for the sudden, fully developed appearance of flowering plants. Island genetics is believed to be a common source of speciation in general, especially when it comes to radical adaptations that seem to have required inferior transitional forms. Flowering plants may have evolved in an isolated setting like an island or island chain, where the plants bearing them were able to develop a highly specialized relationship with some specific animal (a wasp, for example). Such a relationship, with a hypothetical wasp carrying pollen from one plant to another much the way fig wasps do today, could result in the development of a high degree of specialization in both the plant(s) and their partners. The wasp example is not incidental; bees, which, it is postulated, evolved specifically due to mutualistic plant relationships, are descended from wasps.

Animals are also involved in the distribution of seeds. Fruit, which is formed by the enlargement of flower parts, is frequently a seed-dispersal tool that attracts animals to eat or otherwise disturb it, incidentally scattering the seeds it contains. Although many such mutualistic relationships remain too fragile to survive competition and to spread widely, flowering proved to be an unusually effective means of reproduction, spreading (whatever its origin) to become the dominant form of land plant life.

Flower ontogeny uses a combination of genes normally responsible for forming new shoots. The most primitive flowers probably had a variable number of flower parts, often separate from (but in contact with) each other. The flowers tended to grow in a spiral pattern, to be bisexual (in plants, this means both male and female parts on the same flower), and to be dominated by the ovary (female part). As flowers evolved, some variations developed parts fused together, with a much more specific number and design, and with either specific sexes per flower or plant or at least "ovary-inferior".

Flower evolution continues to the present day; modern flowers have been so profoundly influenced by humans that some of them cannot be pollinated in nature. Many modern domesticated flower species were formerly simple weeds, which sprouted only when the ground was disturbed. Some of them tended to grow with human crops, perhaps already having symbiotic companion plant relationships with them, and the prettiest did not get plucked because of their beauty, developing a dependence upon and special adaptation to human affection.

A few paleontologists have also proposed that flowering plants, or angiosperms, might have evolved due to interactions with dinosaurs. One of the idea's strongest proponents is Robert T. Bakker. He proposes that herbivorous dinosaurs, with their eating habits, provided a selective pressure on plants, for which adaptations either succeeded in deterring or coping with predation by herbivores.

Flowering Plant Diversity

A poster of twelve different species of flowers of the *Asteraceae* family

The number of species of flowering plants is estimated to be in the range of 250,000 to 400,000. This compares to around 12,000 species of moss or 11,000 species of pteridophytes, showing that the flowering plants are much more diverse. The number of families in APG (1998) was 462. In APG II (2003) it is not settled; at maximum it is 457, but within this number there are 55 optional segregates, so that the minimum number of families in this system is 402. In APG III (2009) there are 415 families.

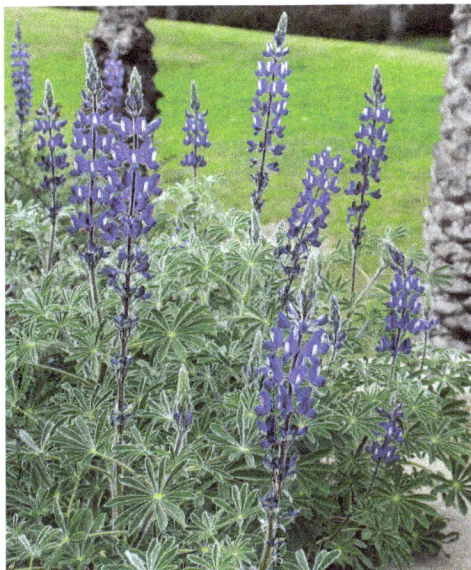

Lupinus pilosus

The diversity of flowering plants is not evenly distributed. Nearly all species belong to the eudicot (75%), monocot (23%), and magnoliid (2%) clades. The remaining 5 clades contain a little over 250 species in total; i.e. less than 0.1% of flowering plant diversity, divided among 9 families.

Ecology

Fertilization and Embryogenesis

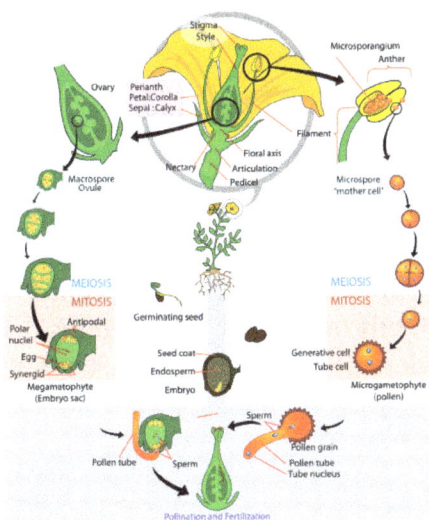

Angiosperm life cycle

Double fertilization refers to a process in which two sperm cells fertilize cells in the ovary. This process begins when a pollen grain adheres to the stigma of the pistil (female reproductive structure), germinates, and grows a long pollen tube. While this pollen tube is growing, a haploid generative cell travels down the tube behind the tube nucleus. The generative cell divides by mitosis to produce two haploid (n) sperm cells. As the pollen tube grows, it makes its way from the stigma, down the style and into the ovary. Here the pollen tube reaches the micropyle of the ovule and digests its way into one of the synergids, releasing its contents (which include the sperm cells). The synergid that the cells were released into degenerates and one sperm makes its way to fertilize the egg cell, producing a diploid ($2n$) zygote. The second sperm cell fuses with both central cell nuclei, producing a triploid ($3n$) cell. As the zygote develops into an embryo, the triploid cell develops into the endosperm, which serves as the embryo's food supply. The ovary will now develop into a fruit and the ovule will develop into a seed.

Fruit and Seed

The fruit of the *Aesculus* or Horse Chestnut tree

As the development of embryo and endosperm proceeds within the embryo sac, the sac wall enlarges and combines with the nucellus (which is likewise enlarging) and the integument to form the *seed coat*. The ovary wall develops to form the fruit or pericarp, whose form is closely associated with the manner of distribution of the seed.

Frequently, the influence of fertilization is felt beyond the ovary, and other parts of the flower take part in the formation of the fruit, e.g., the floral receptacle in the apple, strawberry, and others.

The character of the seed coat bears a definite relation to that of the fruit. They protect the embryo and aid in dissemination; they may also directly promote germination. Among plants with indehiscent fruits, in general, the fruit provides protection for the embryo and secures dissemination. In this case, the seed coat is only slightly developed. If the fruit is dehiscent and the seed is exposed, in general, the seed-coat is well developed, and must discharge the functions otherwise executed by the fruit.

Meiosis

Flowering plants generate gametes using a specialized cell division called meiosis. Meiosis takes place in the ovule (a structure within the ovary that is located within the pistil at the center of the flower). A diploid cell (megaspore mother cell) in the ovule undergoes meiosis (involving two successive cell divisions) to produce four cells (megaspores or female gametes) with haploid nuclei. One of these four cells (megaspore) then undergoes three successive mitotic divisions to produce an immature embryo sac (megagametocyte) with eight haploid nuclei. Next, these nuclei are segregated into separate cells by cytokinesis to producing 3 antipodal cells, 2 synergid cells and an egg cell. Two polar nuclei are left in the central cell of the embryo sac.

Pollen is also produced by meiosis in the male anther (microsporangium). During meiosis, a diploid microspore mother cell undergoes two successive meiotic divisions to produce 4 haploid cells (microspores or male gametes). Each of these microspores, after further mitoses, becomes a pollen grain (microgametophyte) containing two haploid generative (sperm) cells and a tube nucleus. When a pollen grain makes contact with the female stigma, the pollen grain forms a pollen tube that grows down the style into the ovary. In the act of fertilization, a male sperm nucleus fuses with the female egg nucleus to form a diploid zygote that can then develop into an embryo within the newly forming seed. Upon germination of the seed, a new plant can grow and mature.

The adaptive function of meiosis is currently a matter of debate. A key event during meiosis in a diploid cell is the pairing of homologous chromosomes and homologous recombination (the exchange of genetic information) between homologous chromosomes. This process promotes the production of increased genetic diversity among progeny and the recombinational repair of damages in the DNA to be passed on to progeny. To explain the adaptive function of meiosis in flowering plants, some authors emphasize diversity and others emphasize DNA repair

Apomixis

Apomixis (reproduction via asexually formed seeds) is found naturally in about 2.2% of angiosperm genera One type of apomixis, gametophytic apomixis found in a dandelion species involves formation of an unreduced embryo sac due to incomplete meiosis (apomeiosis) and development of an embryo from the unreduced egg inside the embryo sac, without fertilization (parthenogenesis).

Uses

Agriculture is almost entirely dependent on angiosperms, which provide virtually all plant-based food, and also provide a significant amount of livestock feed. Of all the families of plants, the Poaceae, or grass family (grains), is by far the most important, providing the bulk of all feedstocks (rice, corn — maize, wheat, barley, rye, oats, pearl millet, sugar cane, sorghum). The Fabaceae, or legume family, comes in second place. Also of high importance are the Solanaceae, or nightshade family (potatoes, tomatoes, and peppers, among others), the Cucurbitaceae, or gourd family (also including pumpkins and melons), the Brassicaceae, or mustard plant family (including rapeseed and the innumerable varieties of the cabbage species *Brassica oleracea*), and the Apiaceae, or parsley family. Many of our fruits come from the Rutaceae, or rue family (including oranges, lemons, grapefruits, etc.), and the Rosaceae, or rose family (including apples, pears, cherries, apricots, plums, etc.).

In some parts of the world, certain single species assume paramount importance because of their variety of uses, for example the coconut (*Cocos nucifera*) on Pacific atolls, and the olive (*Olea europaea*) in the Mediterranean region.

Flowering plants also provide economic resources in the form of wood, paper, fiber (cotton, flax, and hemp, among others), medicines (digitalis, camphor), decorative and landscaping plants, and many other uses. The main area in which they are surpassed by other plants — namely, coniferous trees (Pinales), which are non-flowering (gymnosperms) — is timber and paper production.

Dicotyledon

The dicotyledons, also known as dicots (or more rarely dicotyls), were one of the two groups into which all the flowering plants or angiosperms were formerly divided. The name refers to one of the typical characteristics of the group, namely that the seed has two embryonic leaves or cotyledons. There are around 200,000 species within this group. The other group of flowering plants were called monocotyledons or monocots, typically having one cotyledon. Historically, these two groups formed the two divisions of the flowering plants.

dicotyledon plant-let

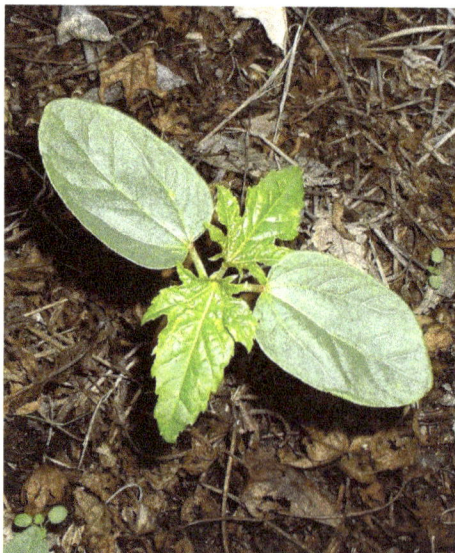

Young castor oil plant showing its prominent two embryonic leaves (cotyledons), that differ from the adult leaves.

Largely from the 1990s onwards, molecular phylogenetic research confirmed what had already been suspected, namely that dicotyledons are not a group made up of all the descendants of a common ancestor (i.e. they are not a monophyletic group). Rather, a number of lineages, such as the magnoliids and groups now collectively known as the basal angiosperms, diverged earlier than the monocots did. The traditional dicots are thus a paraphyletic group. The largest clade of the dicotyledons are known as the eudicots. They are distinguished from all other flowering plants by the structure of their pollen. Other dicotyledons and monocotyledons have monosulcate pollen, or forms derived from it, whereas eudicots have tricolpate pollen, or derived forms, the pollen having three or more pores set in furrows called colpi.

Comparison with Monocotyledons

Aside from cotyledon number, other broad differences have been noted between monocots and dicots, although these have proven to be differences primarily between monocots and eudicots. Many early-diverging dicot groups have "monocot" characteristics such as scattered vascular bundles, trimerous flowers, and non-tricolpate pollen. In addition, some monocots have dicot characteristics such as reticulated leaf veins.

Feature	In monocots	In dicots
Number of parts of each flower	In threes (flowers are trimerous)	In fours or fives (tetramerous or pentamerous)
Number of furrows or pores in pollen	One	Three
Number of cotyledons (leaves in the seed)	One	Two
Arrangement of vascular bundles in the stem	Scattered	In concentric circles
Roots	Are adventitious	Develop from the radicle
Arrangement of major leaf veins	Parallel	Reticulate
Secondary growth	Absent	Often present

Classification

Historical

Traditionally the dicots have been called the Dicotyledones (or Dicotyledoneae), at any rank. If treated as a class, as in the Cronquist system, they could be called the Magnoliopsida after the type genus *Magnolia*. In some schemes, the eudicots were treated as a separate class, the Rosopsida (type genus *Rosa*), or as several separate classes. The remaining dicots (palaeodicots or basal angiosperms) may be kept in a single paraphyletic class, called Magnoliopsida, or further divided. Some botanists prefer to retain the dicotyledons as a valid class, arguing its practicality and that it makes evolutionary sense.

Monocotyledon

Monocotyledons, commonly referred to as monocots, (Lilianae *sensu* Chase & Reveal) are flowering plants (angiosperms) whose seeds typically contain only one embryonic leaf, or cotyledon. They constitute one of the major groups into which the flowering plants have traditionally been divided, the rest of the flowering plants having two cotyledons and therefore classified as dicotyledons, or dicots. However, molecular phylogenetic research has shown that while the monocots form a monophyletic group or clade (comprising all the descendants of a common ancestor), the dicots do not. Monocots have almost always been recognized as a group, but with various taxonomic ranks and under several different names. The APG III system of 2009 recognises a clade called "monocots" but does not assign it to a taxonomic rank.

The monocots include about 60,000 species. The largest family in this group (and in the flowering plants as a whole) by number of species are the orchids (family Orchidaceae), with more than 20,000 species. About half as many species belong to the true grasses (Poaceae), which are economically the most important family of monocots. In agriculture the majority of the biomass produced comes from monocots. These include not only major grains (rice, wheat, maize, etc.), but also forage grasses, sugar cane, and the bamboos. Other economically important monocot crops include various palms (Arecaceae), bananas (Musaceae), gingers and their relatives, turmeric and cardamom (Zingiberaceae), asparagus and the onions and garlic family (Amaryllidaceae). Many houseplants are monocot epiphytes. Additionally most of the horticultural bulbs, plants cultivated for their blooms, such as lilies, daffodils, irises, amaryllis, cannas, bluebells and tulips, are monocots.

Description

Allium crenulatum (Asparagales), an onion, with typical monocot perianth and parallel leaf venation

Onion slice: Parallel veins in cross section

General

The monocots or monocotyledons have, as the name implies, a single (mono-) cotyledon, or embryonic leaf, in their seeds. Historically, this feature was used to contrast the monocots with the dicotyledons or dicots which typically have two cotyledons; however modern research has shown that the dicots are not a natural group, and the term can only be used to indicate all angiosperms that are not monocots and is used in that respect here. From a diagnostic point of view the number of cotyledons is neither a particularly useful characteristic (as they are only present for a very short period in a plant's life), nor is it completely reliable. The single cotyledon is only one of a number of modifications of the body plan of the ancestral monocotyledons, but whose adaptive advantages are poorly understood, but may have been related to adaption to aquatic habitats, prior to radiation to terrestrial habitats. Nevertheless, monocots are sufficiently distinctive that there has rarely been disagreement as to membership of this group, despite considerable diversity in terms of external morphology. However, morphological features that reliably characterise major clades are rare.

Thus monocots are distinguishable from other angiosperms both in terms of their uniformity and diversity. On the one hand the organisation of the shoots, leaf structure and floral configuration are more uniform than in the remaining angiosperms, yet within these constraints a wealth of diversity exists, indicating a high degree of evolutionary success. Monocot diversity includes perennial geophytes such as ornamental flowers including (orchids (Asparagales), tulips and lilies) (Liliales), rosette and succulent epiphytes (Asparagales), mycoheterotrophs (Liliales, Dioscoreales, Pandanales), all in the lilioid monocots, major cereal grains (maize, rice, barley, rye and wheat) in the grass family and forage grasses (Poales) as well as woody tree-like palm trees (Arecales), bamboo, reeds and bromeliads (Poales), bananas and ginger (Zingiberales) in the commelinid monocots, as well as both emergent (Poales, Acorales) and aroids, as well as floating or submerged aquatic plants such as seagrass (Alismatales).

Vegetative

Organisation, Growth and Life Forms

The most important distinction is their growth pattern, lacking a lateral meristem (cambium) that allows for continual growth in diameter with height (secondary growth), and therefore this characteristic is a basic limitation in shoot construction. Although largely herbaceous, some arboraceous monocots reach great height, length and mass. The latter include agaves, palms, pandans, and bamboos. This creates challenges in water transport that monocots deal with in various ways. Some such as species of *Yucca* develop anomalous secondary growth, while palm trees, utilise an anomalous primary growth form described as establishment growth. The axis undergoes primary thickening, that progresses from internode to internode, resulting in a typical inverted conical shape of the basal primary axis. The limited conductivity also contributes to limited branching of the stems. Despite these limitations a wide variety of adaptive growth forms has resulted from epiphytic orchids (Asparagales) and bromeliads (Poales) to submarine Alismatales (including the reduced Lemnoideae) and mycotrophic Burmanniaceae (Dioscreales) and Triuridaceae (Pandanales). Other forms of adaptation include the climbing vines of Araceae (Alismatales) which use negative phototropism (skototropism) to locate host trees (*i.e.* the darkest area), while some palms such as *Calamus manan* (Arecales) produce the longest shoots in the plant kingdom, up to 185 m long. Other monocots, particularly Poales, have adopted a therophyte life form.

Leaves

The cotyledon, the primordial Angiosperm leaf consists of a proximal leaf base or hypophyll and a distal hyperphyll. In moncots the hypophyll tends to be the dominant part in contrast to other angiosperms. From these, considerable diversity arises. Mature monocot leaves are generally narrow and linear, forming a sheathing around the stem at its base, although there are many exceptions. Leaf venation is of the striate type, mainly arcuate-striate or longitunally striate (parallel), less often palmate-striate or pinnate-striate with the leaf veins emerging at the leaf base and then running together at the apices. There is usually only one leaf per node because the leaf base encompasses more than half the circumference. The evolution of this monocot characteristic has been attributed to developmental differences in early zonal differentiation rather than meristem activity (leaf base theory).

Roots and Underground Organs

The lack of cambium in the primary root limits its ability to grow sufficiently to maintain the plant. This necessitates early development of roots derived from the shoot (adventitious roots). In addition to roots, monocots develop runners and rhizomes, which are creeping shoots. Runners serve vegetative propagation, have elongated internodes, run on or just below the surface of the soil and in most case bear scale leaves. Rhizomes frequently have an additional storage function and rhizome producing plants are considered geophytes. Other geophytes develop bulbs, a short axial body bearing leafs whose bases store food. Additional outer non-storage leaves may form a protective function. Other storage organs may be tubers or corms, swollen axes. Tubers may form at the end of underground runners and persist. Corms are short lived vertical shoots with terminal inflorescences and shrivel once flowering has occurred. However, intermediate forms may occur such as in *Crocosmia* (Asparagales). Some monocots may also produce shoots that grow directly down into the soil, these are geophilous shoots that help overcome the limited trunk stability of large woody monocots.

Reproductive

Flowers

In nearly all cases the perigone consists of two alternating trimerous whorls of tepals, being homochlamydeous, without differentiation between calyx and corolla. In zoophilous (pollinated by animals) taxa, both whorls are corolline (petal-like). Anthesis (the period of flower opening) is usually short fugacious (short lived). Some of the more persistent perigones demonstrate thermonastic opening and closing (responsive to changes in temperature). About two thirds of monocots are zoophilous, predominantly by insects. These plants need to advertise to pollinators and do so by way of phaneranthous (showy) flowers. Such optical signalling is usually a function of the tepal whorls but may also be provided by semaphylls (other structures such as filaments, staminodes or stylodia which have become modified to attract pollinators). However, some monocot plants may have aphananthous (inconspicuous) flowers and still be pollinated by animals. In these the plants rely either on chemical attraction or other structures such as coloured bracts fulfill the role of optical attraction. In some phaneranthous plants such structures may reinforce floral structures.

The production of fragrances for olfactory signalling are common in monocots. The perigone also functions as a landing platform for pollinating insects.

Fruit and Seed

The embryo consists of a single cotyledon, usually with two vascular bundles.

Comparison with "Dicots"

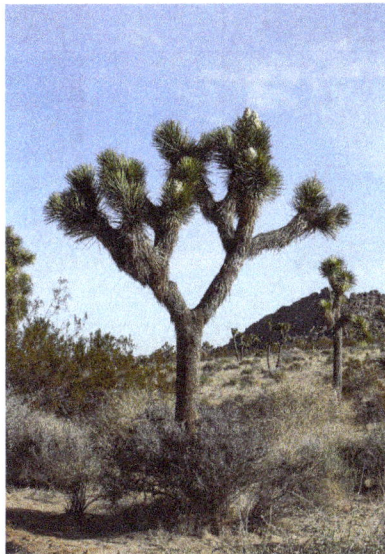

Yucca brevifolia (Joshua Tree: Asparagales)

The traditionally listed differences between monocots and "dicots" are as follows. This is a broad sketch only, not invariably applicable, as there are a number of exceptions. The differences indicated are more true for monocots versus eudicots.

Feature	In monocots	In "dicots"
Growth form	Mostly herbaceous, occasionally arboraceous	Herbaceous or arboraceous
Leaves	Leaf shape oblong or linear, often sheathed at base, petiole seldom developed, stipules absent. Major leaf veins usually parallel	Broad, seldom sheathed, petiole common often with stipules. Veins usually reticulate (pinnate or palmate)
Roots	Primary root of short duration, replaced by adventitial roots forming fibrous or fleshy root systems	Develops from the radicle. Primary root often persists forming strong taproot and secondary roots
Plant stem: Vascular bundles	Numerous scattered bundles in ground parenchyma, cambium rarely present, no differentiation between cortical and stelar regions	Ring of primary bundles with cambium, differentiated into cortex and stele (eustelic)
Flowers	Parts in threes (trimerous) or multiples of three (*e.g.* 3, 6 or 9 petals)	Fours (tetramerous) or fives (pentamerous)
Pollen: Number of apertures (furrows or pores)	Monocolpate (single aperture or colpus)	Tricolpate (three)
Embryo: Number of cotyledons (leaves in the seed)	One, endosperm frequently present in seed	Two, endosperm present or absent

Comparison of Monocots and "Dicots"

	MONOCOT		DICOT	
Single Cotyledon		Two Cotyledon		
Long Narrow Leaf Parallel Veins		Broad Leaf Network of Veins		
Vascular Bundles Scattered		Vascular Bundles in a Ring		
Floral Parts in Multiples of 3		Floral Parts in Multiples of 4 or 5		

A number of these differences are not unique to the monocots, and while still useful no one single feature, will infallibly identify a plant as a monocot. For example, trimerous flowers and monosulcate pollen are also found in magnoliids, of which exclusively adventitious roots are found in some of the Piperaceae. Similarly, at least one of these traits, parallel leaf veins, is far from universal among the monocots. Monocots with broad leaves and reticulate leaf veins, typical of dicots, are found in a wide variety of monocot families: for example, *Trillium*, *Smilax* (greenbriar), and *Pogonia* (an orchid), and the Dioscoreales (yams). *Potamogeton* are one of several monocots with tetramerous flowers. Other plants exhibit a mixture of characteristics. Nymphaeaceae (water lilies) have reticulate veins, a single cotyledon, adventitious roots and a monocot like vascular bundle. These examples reflect their shared ancestry. Nevertheless, this list of traits is a generally valid set of contrasts, especially when contrasting monocots with eudicots rather than non-monocot flowering plants in general.

Apomorphies

Monocot apomorphies (characteristics that are derived during radiation rather than inherited from an ancestral form) include herbaceous habit, leaves with parallel venation and sheathed base, embryo with a single cotyledon, atactostele stele, numerous adventitious roots, sympodial growth, and trimerous (3 parts per whorl) flowers that are pentacyclic (5 whorled) with 3 sepals, 3 petals, 2 whorls of 3 stamens each and 3 carpels. In contrast monosculate pollen is considered an ancestral trait, probably plesiomorphic.

Synapomorphies

The distinctive features of the monocots have contributed to the relative taxonomic stability of the group. Douglas E. Soltis and others identify thirteen synapomorphies (shared characteristics that unite monophyletic groups of taxa):

1. Calcium oxalate raphides

2. Absence of vessels in leaves

3. Monocotyledonous anther wall formation

4. Successive microsporogenesis

5. Syncarpous gynoecium

6. Parietal placentation

7. Monocotyledonous seedling

8. Persistent radicle

9. Haustorial cotyledon tip

10. Open cotyledon sheath

11. Steroidal saponins

12. Fly pollination

13. Diffuse vascular bundles and absence of secondary growth

Vascular System

Monocots have a distinctive arrangement of vascular tissue known as an atactostele in which the vascular tissue is scattered rather than arranged in concentric rings. Collenchyma is absent in monocot stems, roots and leaves. Many monocots are herbaceous and do not have the ability to increase the width of a stem (secondary growth) via the same kind of vascular cambium found in non-monocot woody plants. However, some monocots do have secondary growth, and because it does not arise from a single vascular cambium producing xylem inwards and phloem outwards, it is termed "anomalous secondary growth". Examples of large monocots which either exhibit secondary growth, or can reach large sizes without it, are palms (Arecaceae), screwpines (Pandanaceae), bananas (Musaceae), *Yucca, Aloe, Dracaena,* and *Cordyline.*

Roystonea regia palm (Arecales) stems showing anomalous secondary growth in monocots, with characteristic fibrous roots

Taxonomy

The monocots form one of five major lineages of mesangiosperms (core angiosperms), which in themselves form 99.95% of all angiosperms. The monocots and the eudicots, are the largest and most diversified angiosperm radiations accounting for 22% and 75% of all angiosperm species respectively.

Of these, the grass family (Poaceae) is the most economically important, which together with the orchids Orchidaceae account for half of the species diversity, accounting for 34% and 17% of all monocots respectively and are among the largest families of angiosperms. They are also among the dominant members of many plant communities.

Early History

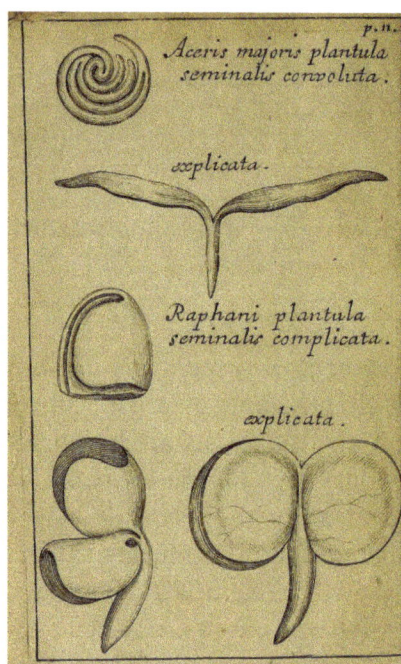

Illustrations of cotyledons by John Ray 1682, after Malpighi

The monocots are one of the major divisions of the flowering plants or angiosperms. They have been recognized as a natural group since John Ray's studies of seed structure in the 17th century. Ray was the first botanical systematist, and in his examination of seeds, first observed the dichotomy of cotyledon structure. He reported his findings in a paper read to the Royal Society on 17 December 1674, entitled "A Discourse on the Seeds of Plants".

Since this paper appeared a year before the publication of Malpighi's *Anatome Plantarum* (1675–1679), Ray has the priority. At the time, Ray did not fully realise the importance of his discovery but progressively developed this over successive publications. And since these were in Latin, "seed leaves" became *folia seminalia* and then *cotyledon*, following Malpighi. Malpighi and Ray were familiar with each other's work, and Malpighi in describing the same structures had introduced the term cotyledon, which Ray adopted in his subsequent writing.

In this experiment, Malpighi also showed that the cotyledons were critical to the development of the plant, proof that Ray required for his theory. In his *Methodus plantarum nova* Ray also developed and justified the "natural" or pre-evolutionary approach to classification, based on characteristics selected *a posteriori* in order to group together taxa that have the greatest number of shared characteristics. This approach, also referred to as polythetic would last till evolutionary theory enabled Eichler to develop the phyletic system that superseded it in the late nineteenth century, based on an understanding of the acquisition of characteristics. He also made the crucial observation *Ex hac seminum divisione sumum potest generalis plantarum distinctio, eaque meo judicio omnium prima et longe optima, in eas sci. quae plantula seminali sunt bifolia aut διλόβω, et quae plantula sem. adulta analoga.* (From this division of the seeds derives a general distinction amongst plants, that in my judgement is first and by far the best, into those seed plants which are bifoliate, or bilobed, and those that are analogous to the adult), that is between monocots and dicots. He illustrated this with by quoting from Malpighi and including reproductions of Malpighi's drawings of cotyledons. Initially Ray did not develop a classification of flowering plants (florifera) based on a division by the number of cotyledons, but developed his ideas over successive publications, coining the terms *Monocotyledones* and *Dicotyledones* in 1703, in the revised version of his *Methodus* (*Methodus plantarum emendata*), as a primary method for dividing them, *Herbae floriferae, dividi possunt, ut diximus, in Monocotyledones & Dicotyledones* (Flowering plants, can be divided, as we have said, into Monocotyledons & Dicotyledons).

Post Linnean

Although Linnaeus (1707–1778) did not utilise Ray's discovery, basing his own classification solely on floral reproductive morphology, the term was used shortly after his classification appeared (1753) by Scopoli and who is credited for its introduction. Every taxonomist since then, starting with De Jussieu and De Candolle, has used Ray's distinction as a major classification characteristic. In De Jussieu's system (1789), he followed Ray, arranging his Monocotyledones into three classes based on stamen position and placing them between Acotyledones and Dicotyledones. De Candolle's system (1813) which was to predominate thinking through much of the 19th century used a similar general arrangement, with two subgroups of his *Monocotylédonés* (Monocotyledoneae). Lindley (1830) followed De Candolle in using the terms Monocotyledon and Endogenae interchangeably. They considered the monocotyledons to be a group of vascular plants (*Vasculares*) whose vascular bundles were thought to arise from within (*Endogènes* or endogenous).

Monocotyledons remained in a similar position as a major division of the flowering plants throughout the nineteenth century, with minor variations. George Bentham and Hooker (1862–1883) used Monocotyledones, as would Wettstein, while August Eichler used Mononocotyleae and Engler, following de Candolle, Monocotyledoneae. In the twentieth century, some authors used alternative names such as Bessey's (1915) Alternifoliae and Cronquist's (1966) Liliatae. Later (1981) Cronquist changed Liliatae to Liliopsida, usages also adopted by Takhtajan simultaneously. Thorne (1992) and Dahlgren (1985) also used Liliidae as a synonym.

Taxonomists had considerable latitude in naming this group, as the Monocotyledons were a group above the rank of family. Article 16 of the *ICBN* allows either a descriptive name or a name formed from the name of an included family.

In summary they have been variously named, as follows:

- class Monocotyledoneae in the de Candolle system and the Engler system
- class Monocotyledones in the Bentham & Hooker system and the Wettstein system

- class Monocotyleae in the Eichler system

- class Liliatae then Liliopsida in the Takhtajan system and the Cronquist system

- subclass Liliidae in the Dahlgren system and the Thorne system

Modern Era

Over the 1980s, a more general review of the classification of angiosperms was undertaken. The 1990s saw considerable progress in plant phylogenetics and cladistic theory, initially based on *rbcL* gene sequencing and cladistic analysis, enabling a phylogenetic tree to be constructed for the flowering plants. The establishment of major new clades necessitated a departure from the older but widely used classifications such as Cronquist and Thorne, based largely on morphology rather than genetic data. These developments complicated discussions on plant evolution and necessitated a major taxonomic restructuring.

This DNA based molecular phylogenetic research confirmed on the one hand that the monocots remained as a well defined monophyletic group or clade, in contrast to the other historical divisions of the flowering plants, which had to be substantially reorganized. No longer could the angiosperms be simply divided into monocotyledons, and dicotyledons but it was apparent that the monocotyledons were but one of a relatively large number of defined groups within the angiosperms. Correlation with morphological criteria showed that the defining feature was not cotyledon number but the separation of angiosperms into two major pollen types, uniaperturate (monosulcate and monosulcate-derived) and triaperturate (tricolpate and tricolpate-derived), with the monocots situated within the uniaperturate groups. The formal taxonomic ranking of Monoctyledons thus became replaced with monocots as an informal clade. This is the name that has been most commonly used since the publication of the Angiosperm Phylogeny Group (APG) system in 1998 and regularly updated since.

Within the angiosperms, there are two major grades, a small early branching basal grade, the basal angiosperms (ANA grade) with three lineages and a larger late branching grade, the core angiosperms (mesangiosperms) with five lineages, as shown in the cladogram.

Subdivision

While the monocotyledons have remained extremely stable in their outer borders as a well-defined and coherent monophylectic group, the deeper internal relationships have undergone considerable flux, with many competing classification systems over time.

Historically, Bentham (1877), considered the monocots to consist of four alliances, Epigynae, Coronariae, Nudiflorae and Glumales, based on floral characteristics. He describes the attempts to subdivide the group since the days of Lindley as largely unsuccessful. Like most subsequent classification systems it failed to distinguish between two major orders, Liliales and Asparagales, now recognised as quite separate. A major advance in this respect was the work of Rolf Dahlgren (1980), which would form the basis of the Angiosperm Phylogeny Group's (APG) subsequent modern classification of monocot families. Dahlgren who used the alternate name Lilliidae considered the monocots as a subclass of angiosperms characterised by a single cotyledon and the presence of triangular protein bodies in the sieve tube plastids. He divided the monocots into seven superorders, Alismatiflorae, Ariflorae, Triuridiflorae, Liliiflorae, Zingiberiflorae, Commeliniflorae and Areciflorae. With respect to the specific issue regarding Liliales and Asparagales, Dahlgren followed Huber (1969) in adopting a splitter approach, in contrast to the longstanding tendency to view Liliaceae as a very broad sensu lato family. Following Dahlgren's untimely death in 1987, his work was continued by his widow, Gertrud Dahlgren, who published a revised version of the classification in 1989. In this scheme the suffix -*florae* was replaced with -*anae* (*e.g.* Alismatanae) and the number of superorders expanded to ten with the addition of Bromelianae, Cyclanthanae and Pandananae.

Molecular studies have both confirmed the monophyly of the monocots and helped elucidate relationships within this group. The APG system does not assign the monocots to a taxonomic rank, instead recognizing a monocots clade. However, there has remained some uncertainty regarding the exact relationships between the major lineages, with a number of competing models (including APG).

The APG system establishes eleven orders of monocots. These form three grades, the alismatid monocots, lilioid monocots and the commelinid monocots by order of branching, from early to late. In the following cladogram numbers indicate crown group (most recent common ancestor of the sampled species of the clade of interest) divergence times in mya (million years ago).

Of some 70,000 species, by far the largest number (65%) are found in two families, the orchids and grasses. The orchids (Orchidaceae, Asparagales) contain about 25,000 species and the grasses (Poaceae, Poales) about 11,000. Other well known groups within the Poales order include the Cyperaceae (sedges) and Juncaceae (rushes), and the monocots also include familiar families such as the palms (Arecaceae, Arecales) and lilies (Liliaceae, Liliales).

Evolution

In prephyletic classification systems monocots were generally positioned between plants other than angiosperms and dicots, implying that monocots were more primitive. With the introduction of phyletic thinking in taxonomy (from the system of Eichler 1875–1878 onwards) the predominant theory of monocot origins was the ranalean (ranalian) theory, particularly in the work of Bessey (1915), which traced the origin of all flowering plants to a Ranalean type, and reversed the sequence making dicots the more primitive group.

The monocots form a monophyletic group arising early in the history of the flowering plants, but the fossil record is meagre. The earliest fossils presumed to be monocot remains date from the early Cretaceous period. For a very long time, fossils of palm trees were believed to be the oldest monocots, first appearing 90 million years ago (mya), but this estimate may not be entirely true. At least some putative monocot fossils have been found in strata as old as the eudicots. The oldest fossils that are unequivocally monocots are pollen from the Late Barremian–Aptian – Early Cretaceous period, about 120-110 million years ago, and are assignable to clade-Pothoideae-Monstereae Araceae; being Araceae, sister to other Alismatales. They have also found flower fossils of Triuridaceae (Pandanales) in Upper Cretaceous rocks in New Jersey, becoming the oldest known sighting of saprophytic/mycotrophic habits in angiosperm plants and among the oldest known fossils of monocotyledons.

Topology of the angiosperm phylogenetic tree could infer that the monocots would be among the oldest lineages of angiosperms, which would support the theory that they are just as old as the eudicots. The pollen of the eudicots dates back 125 million years, so the lineage of monocots should be that old too.

Molecular Clock Estimates

Kåre Bremer, using rbcL sequences and the mean path length method for estimating divergence times, estimated the age of the monocot crown group (i.e. the time at which the ancestor of today's *Acorus* diverged from the rest of the group) as 134 million years. Similarly, Wikström *et al.*, using Sanderson's non-parametric rate smoothing approach, obtained ages of 127–141 million years for the crown group of monocots. All these estimates have large error ranges (usually 15-20%), and Wikström *et al.* used only a single calibration point, namely the split between Fagales and Cucurbitales, which was set to 84 Ma, in the late Santonian period. Early molecular clock studies using strict clock models had estimated the monocot crown age to 200 ± 20 million years ago or 160 ± 16 million years, while studies using relaxed clocks have obtained 135-131 million years or 133.8 to 124 million years. Bremer's estimate of 134 million years has been used as a secondary calibration point in other analyses. Some estimates place the emergence of the monocots as far back as 150 mya in the Jurassic period.

Core Group

The age of the core group of so-called 'nuclear monocots' or 'core monocots', which correspond to all orders except Acorales and Alismatales, is about 131 million years to present, and crown group age is about 126 million years to the present. The subsequent branching in this part of the tree (i.e. Petrosaviaceae, Dioscoreales + Pandanales and Liliales clades appeared), including the crown Petrosaviaceae group may be in the period around 125–120 million years BC (about 111 million years so far), and stem groups of all other orders, including Commelinidae would have diverged about or shortly after 115 million years. These and many clades within these orders may have originated in southern Gondwana, i.e. Antarctica, Australasia, and southern South America.

Aquatic Monocots

The aquatic monocots of Alismatales have commonly been regarded as "primitive". They have also been considered to have the most primitive foliage, which were cross-linked as Dioscoreales and Melanthiales. Keep in mind that the "most primitive" monocot is not necessarily "the sister of everyone else". This is because the ancestral or primitive characters are inferred by means of the reconstruction of character states, with the help of the phylogenetic tree. So primitive characters of monocots may be present in some derived groups. On the other hand, the basal taxa may exhibit many morphological autapomorphies. So although Acoraceae is the sister group to the remaining monocotyledons, the result does not imply that Acoraceae is "the most primitive monocot" in terms of its character states. In fact, Acoraceae is highly derived in many morphological characters, and that is precisely why Acoraceae and Alismatales occupied relatively derived positions in the trees produced by Chase *et al.* and others.

Some authors support the idea of an aquatic phase as the origin of monocots. The phylogenetic position of Alismatales (many water), which occupy a relationship with the rest except the Acoraceae, do not rule out the idea, because it could be 'the most primitive monocots' but not 'the most basal'. The Atactostele stem, the long and linear leaves, the absence of secondary growth, roots in groups instead of a single root branching (related to the nature of the substrate), including sympodial use, are consistent with a water source. However, while monocots were sisters of the aquatic Ceratophyllales, or their origin is related to the adoption of some form of aquatic habit, it would not help much to the understanding of how it evolved to develop their distinctive anatomical features: the monocots seem so different from the rest of angiosperms and it's difficult to relate their morphology, anatomy and development and those of broad-leaved angiosperms.

Other Taxa

In the past, taxa which had petiolate leaves with reticulate venation were considered "primitive" within the monocots, because of its superficial resemblance to the leaves of dicotyledons. Recent work suggests that these taxa are sparse in the phylogenetic tree of monocots, such as fleshy fruited taxa (excluding taxa with aril seeds dispersed by ants), the two features would be adapted to conditions that evolved together regardless. Among the taxa involved were *Smilax*, *Trillium* (Liliales), *Dioscorea* (Dioscoreales), etc. A number of these plants are vines that tend to live in shaded habitats for at least part of their lives, and may also have a relationship with their shapeless stomata. Reticulate venation seems to have appeared at least 26 times in monocots, in fleshy fruits 21 times (sometimes lost later), and the two characteristics, though different, showed strong signs of a tendency to be good or bad in tandem, a phenomenon described as "concerted convergence" ("coordinated convergence").

Etymology

The name monocotyledons is derived from the traditional botanical name "Monocotyledones" or *Monocotyledoneae* in Latin, which refers to the fact that most members of this group have one cotyledon, or embryonic leaf, in their seeds.

Ecology

Emergence

Some monocots, such as grasses, have hypogeal emergence, where the mesocotyl elongates and pushes the coleoptile (which encloses and protects the shoot tip) toward the soil surface. Since elongation occurs above the cotyledon, it is left in place in the soil where it was planted. Many dicots have epigeal emergence, in which the hypocotyl elongates and becomes arched in the soil. As the hypocotyl continues to elongate, it pulls the cotyledons upward, above the soil surface.

Conservation

The IUCN Red List describes four species as extinct, four as extinct in the wild, 626 as possibly extinct, 423 as critically endangered, 632 endangered, 621 vulnerable, and 269 near threatened of 4,492 whose status is known.

Uses

Monocots are among the most important plants, economically and culturally and account for most of the staple foods of the world, such as cereal grains and starchy root crops, palms, orchids and lilies, building materials, and many medicines. Of the monocots, the grasses are of enormous economic importance as a source of animal and human food, and form the largest component of agricultural species in terms of biomass produced.

Araceae

Snake lily (*Dracunculus vulgaris*) in Crete

The Araceae are a family of monocotyledonous flowering plants in which flowers are borne on a type of inflorescence called a spadix. The spadix is usually accompanied by, and sometimes partially enclosed in, a spathe or leaf-like bract. Also known as the arum family, members are often colloquially known as aroids. This family of 114 genera and about 3750 known species is most diverse in the New World tropics, although also distributed in the Old World tropics and northern temperate regions.

Peace lily (*Spathiphyllum cochlearispathum*) clearly showing the
characteristic spadix and spathe

The largest collection of living Araceae is maintained at the Missouri Botanical Gardens. Another large collection of living Araceae can be found at the Munich Botanical Garden, due to the efforts of researcher and aroid authority Josef Bogner.

Description

Species in the Araceae are often rhizomatous or tuberous and are often found to contain calcium oxalate crystals or raphides. The leaves can vary considerably from species to species. The inflorescence is composed of a spadix, which is almost always surrounded by a modified leaf called a spathe. In monoecious aroids (possessing separate male and female flowers, but with both flowers present on one plant), the spadix is usually organized with female flowers towards the bottom and male flowers towards the top. In aroids with perfect flowers, the stigma is no longer receptive when the pollen is released, thus preventing self-fertilization. Some species are dioecious.

Many plants in this family are thermogenic (heat-producing). Their flowers can reach up to 45 °C even when the surrounding air temperature is much lower. One reason for this unusually high temperature is to attract insects (usually beetles) to pollinate the plant, rewarding the beetles with heat energy. Another reason is to prevent tissue damage in cold regions. Some examples of thermogenic Araceae are: *Symplocarpus foetidus* (eastern skunk cabbage), *Amorphophallus titanum* (titan arum), *Amorphophallus paeoniifolius* (elephant foot yam), *Helicodiceros muscivorus* (dead horse arum lily), and *Sauromatum venosum* (voodoo lily). Species such as titan arum and the dead horse arum give off a very pungent smell, often resembling rotten flesh, to attract flies to pollinate the plant. The heat produced by the plant helps to convey the scent further.

Taxonomy

Classification

One of the earliest observations of species in the Araceae was conducted by Theophrastus in his work *Enquiry into Plants*. The Araceae were not recognized as a distinct group of plants until the 16th century. In 1789, Antoine Laurent de Jussieu classified all climbing aroids as *Pothos* and all terrestrial aroids as either *Arum* or *Dracontium* in his book *Familles des Plantes*.

The first major system of classification for the family was produced by Heinrich Wilhelm Schott, who published *Genera Aroidearum* in 1858 and *Prodromus Systematis Aroidearum* in 1860. Schott's system was based on floral characteristics, and used a narrow conception of a genus. Adolf Engler produced a classification in 1876, which was steadily refined up to 1920. His system is significantly different from Schott's, being based more on vegetative characters and anatomy. The two systems were to some extent rivals, with Engler's having more adherents before the advent of molecular phylogenetics brought new approaches.

Modern studies based on gene sequences show the Araceae (including the Lemnoideae, duckweeds) to be monophyletic, and the first diverging group within the Alismatales. The APG III system of 2009 recognizes the family, including the genera formerly segregated in the Lemnaceae. The sinking of the Lemnaceae into the Araceae is not universally accepted. For example, the 2010

New Flora of the British Isles uses a paraphyletic Araceae and a separate Lemnaceae. A comprehensive genomic study of *Spirodela polyrhiza* was published in February 2014.

Genera

Anthurium and *Zantedeschia* are two well-known members of this family, as are *Colocasia esculenta* (taro) and *Xanthosoma roseum* (elephant ear or 'ape). The largest unbranched inflorescence in the world is that of the arum *Amorphophallus titanum* (titan arum). The family includes many ornamental plants: *Dieffenbachia*, *Aglaonema*, *Caladium*, *Nephthytis*, and *Epipremnum*, to name a few. In the genus *Cryptocoryne* are many popular aquarium plants. *Philodendron* is an important plant in the ecosystems of the rainforests and is often used in home and interior decorating. *Symplocarpus foetidus* (skunk cabbage) is a common eastern North American species. An interesting peculiarity is that this family includes the largest unbranched inflorescence, that of the titan arum, often erroneously called the "largest flower" and the smallest flowering plant and smallest fruit, found in the duckweed, *Wolffia*.

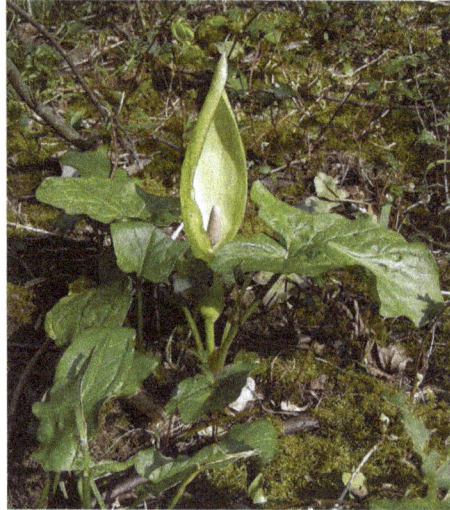

The cuckoo-pint or lords and ladies (*Arum maculatum*) is a common arum in British woodlands.

Arisaema triphyllum

Toxicity

Within the Aracae, genera such as *Alocasia, Arisaema, Caladium, Colocasia, Dieffenbachia*, and *Philodendron* contain calcium oxalate crystals in the form of raphides. When consumed, these may cause edema, vesicle formation, and dysphagia accompanied by painful stinging and burning to the mouth and throat, with symptoms occurring for up to two weeks after ingestion.

Food Plants

Food plants in the Araceae include *Colocasia esculenta* (taro, dasheen), *Xanthosoma* (cocoyam, tannia), and *Monstera deliciosa* (Mexican breadfruit). While the aroids are little traded, and overlooked by plant breeders to the extent that the Crop Trust calls them "orphan crops", they are widely grown and are important in subsistence agriculture and in local markets. The main food product is the corm, which is high in starch; leaves and flowers also find culinary use.

Iridaceae

Iridaceae is a family of plants in order Asparagales, taking its name from the irises, meaning rainbow, referring to its many colours. There are 66 accepted genera with a total of c. 2244 species worldwide (Christenhusz & Byng 2016). It includes a number of other well known cultivated plants, such as freesias, gladioli and crocuses.

Members of this family are perennial plants, with a bulb, corm or rhizome. The plants grow erect, and have leaves that are generally grass-like, with a sharp central fold. Some examples of members of this family are the blue flag and yellow flag.

Name and History

The family name is based on the genus Iris, the largest and best known genus in Europe. The genus Iris dates from 1753, when it was coined by Swedish botanist, Carl Linnaeus. Its name derives from the Greek goddess, Iris, who carried messages from Olympus to earth along a rainbow, whose colours were seen by Linnaeus in the multi-hued petals of many of the species.

The family is currently divided into four subfamilies but the results from DNA analysis suggest that several more should be recognised:

Subfamily Crocoideae is one of the major subfamilies in the Iridaceae family, it contains many genera, including Afrocrocus, Babiana, Chasmanthe, Crocosmia, Crocus, Cyanixia, Devia, Dierama, Duthieastrum, Freesia, Geissorhiza, Gladiolus, Hesperantha, Ixia, Lapeirousia, Melasphaerula, Micranthus, Pillansia, Romulea, Sparaxis, Savannosiphon, Syringodea, Thereianthus, Tritonia, Tritoniopsis, Xenoscapa and Watsonia. They are mainly from Africa, but includes members from Europe and Asia. The rootstock is usually a corm, they have blooms which sometimes have scent are collected in inflorescence and contain six tepals. The nectar is produced mostly in the base of the bloom from the glands of the ovary, which is where the flower forms a tube-like end. In some species there is no such end and the plant only provides pollen to pollinating insects. Like the whole Iridaceae family, the members of the subfamily have the typical sword-shaped leaves.

Subfamily Isophysidoideae contains the single genus Isophysis, from Tasmania. It is the only member of the family with a superior ovary and has a star-like yellow to brownish flower.

Subfamily Nivenioideae contains six genera from South Africa, Australia and Madagascar, including the only true shrubs in the family (Klattia, Nivenia and Witsenia) as well as the only myco-heterotroph (Geosiris). Aristea is also a member of this subfamily. It is distinguished by having flowers in small, paired clusters among large bracts, slender styles that are divided into three slender branches and nectar (when present) produced from glands in the ovary walls. The flowers are always radially symmetrical, with separate tepals (petals) and the rootstock is a rhizome.

Subfamily Iridoideae is distributed throughout the range of the family and contains the large genera Iris and Moraea. It is the only subfamily that is represented in South America. The species have flowers in solitary clusters among large bracts, styles that are often petal-like or crested and nectar (when present) is produced from glands on the tepals. Most species have separate petals and the rootstock is usually a rhizome or rarely a bulb. The flowers are almost always radially symmetrical. Bobartia, Dietes and Ferraria belong to this subfamily.

Ecology

Members of Iridaceae occur in a great variety of habitats. About the only place they do not grow is in the sea itself, although Gladiolus gueinzii occurs on the seashore just above the high tide mark within reach of the spray. Most species are adapted to seasonal climates that have a pronounced dry or cold period unfavourable for plant growth and during which the plants dormant. As a result, most species are deciduous. Evergreen species are restricted to subtropical forests or savannah, temperate grasslands and perennially moist fynbos. A few species grow in marshes or along streams and some even grow only in the spray of seasonal waterfalls.

The above ground parts (leaves and stems) of deciduous species die down when the bulb or corm enters dormancy. The plants thus survive periods that are unfavourable for growth by retreating underground. This is particularly useful in grasslands and fynbos, which are adapted to regular burning in the dry season. At this time the plants are dormant and their bulbs or corms are able to survive the heat of the fires underground. Veld fires clear the soil surface of competing vegetation, as well as fertilise it with ash. With the arrival of the first rains, the dormant corms are ready to burst into growth, sending up flowers and stems before they can be shaded out by other vegetation. Many grassland and fynbos irids flower best after fires and some fynbos species will only flower in the season after a fire.

The family has a very diverse pollination ecology. Most species are pollinated by various species of solitary bees but many are adapted to pollination by sunbirds. These species typically have red to orange, trumpet-like flowers that secrete large amounts of nectar. Other species are adapted to pollination by butterflies and moths, carrion flies and long-proboscid flies, and even monkey-beetles.

Orchidaceae

The Orchidaceae are a diverse and widespread family of flowering plants, with blooms that are often colourful and often fragrant, commonly known as the orchid family.

Along with the Asteraceae, they are one of the two largest families of flowering plants. The Orchidaceae have about 28,000 currently accepted species, distributed in about 763 genera. The determination of which family is larger is still under debate, because verified data on the members of such enormous families are continually in flux. Regardless, the number of orchid species nearly equals the number of bony fishes and is more than twice the number of bird species, and about four times the number of mammal species. The family also encompasses about 6–11% of all seed plants. The largest genera are *Bulbophyllum* (2,000 species), *Epidendrum* (1,500 species), *Dendrobium* (1,400 species) and *Pleurothallis* (1,000 species).

The family also includes *Vanilla* (the genus of the vanilla plant), *Orchis* (type genus), and many commonly cultivated plants such as *Phalaenopsis* and *Cattleya*. Moreover, since the introduction of tropical species into cultivation in the 19th century, horticulturists have produced more than 100,000 hybrids and cultivars.

Description

High resolution image of orchid

Orchids are easily distinguished from other plants, as they share some very evident, shared derived characteristics, or "apomorphies". Among these are: bilateral symmetry of the flower (zygomorphism), many resupinate flowers, a nearly always highly modified petal (labellum), fused stamens and carpels, and extremely small seeds.

Stem and Roots

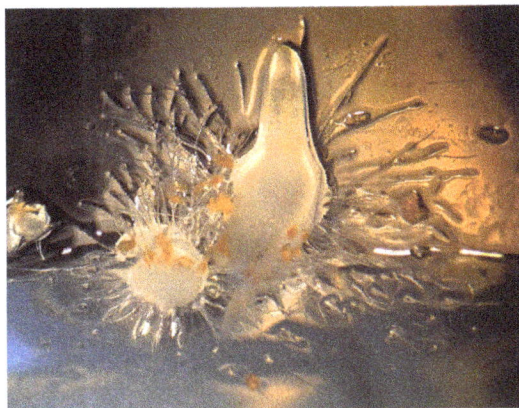

Germinating seeds of the temperate orchid *Anacamptis coriophora*. The protocorm is the first organ that will develop into true roots and leaves.

All orchids are perennial herbs that lack any permanent woody structure. They can grow according to two patterns:

- Monopodial: The stem grows from a single bud, leaves are added from the apex each year and the stem grows longer accordingly. The stem of orchids with a monopodial growth can reach several metres in length, as in *Vanda* and *Vanilla*.

- Sympodial: Sympodial orchids have a front (the newest growth) and a back (the oldest growth). The plant produces a series of adjacent shoots which grow to a certain size, bloom and then stop growing and are replaced. Sympodial orchids grow laterally rather than vertically, following the surface of their support. The growth continues by development of new leads, with their own leaves and roots, sprouting from or next to those of the previous year, as in *Cattleya*. While a new lead is developing, the rhizome may start its growth again from a so-called 'eye', an undeveloped bud, thereby branching. Sympodial orchids may have visible pseudobulbs joined by a *rhizome*, which creeps along the top or just beneath the soil.

Anacamptis lactea showing the two tubers

Terrestrial orchids may be rhizomatous or form corms or tubers. The root caps of terrestrial orchids are smooth and white.

Some sympodial terrestrial orchids, such as *Orchis* and *Ophrys*, have two subterranean tuberous roots. One is used as a food reserve for wintry periods, and provides for the development of the other one, from which visible growth develops.

In warm and constantly humid climates, many terrestrial orchids do not need pseudobulbs.

Epiphytic orchids, those that grow upon a support, have modified aerial roots that can sometimes be a few meters long. In the older parts of the roots, a modified spongy epidermis, called velamen, has the function to absorb humidity. It is made of dead cells and can have a silvery-grey, white or brown appearance. In some orchids, the velamen includes spongy and fibrous bodies near the passage cells, called tilosomes.

The cells of the root epidermis grow at a right angle to the axis of the root to allow them to get a firm grasp on their support. Nutrients for epiphytic orchids mainly come from mineral dust, organic detritus, animal droppings and other substances collecting among on their supporting surfaces.

The base of the stem of sympodial epiphytes, or in some species essentially the entire stem, may be thickened to form a pseudobulb that contains nutrients and water for drier periods.

The pseudobulb has a smooth surface with lengthwise grooves, and can have different shapes, often conical or oblong. Its size is very variable; in some small species of *Bulbophyllum*, it is no longer than two millimeters, while in the largest orchid in the world, *Grammatophyllum speciosum* (giant orchid), it can reach three meters. Some *Dendrobium* species have long, canelike pseudobulbs with short, rounded leaves over the whole length; some other orchids have hidden or extremely small pseudobulbs, completely included inside the leaves.

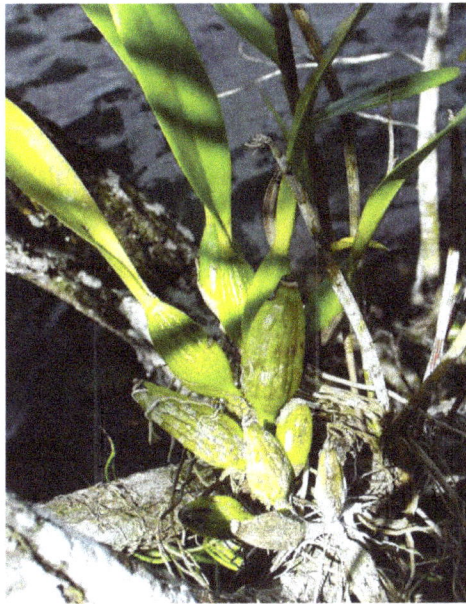

The pseudobulb of *Prosthechea fragrans*

With ageing, the pseudobulb sheds its leaves and becomes dormant. At this stage, it is often called a backbulb. Backbulbs still hold nutrition for the plant, but then a pseudobulb usually takes over, exploiting the last reserves accumulated in the backbulb, which eventually dies off, too. A pseudobulb typically lives for about five years. Orchids without noticeable pseudobulbs are also said to have growths, an individual component of a sympodial plant.

Leaves

Like most monocots, orchids generally have simple leaves with parallel veins, although some Vanilloideae have reticulate venation. Leaves may be ovate, lanceolate, or orbiculate, and very variable in size on the individual plant. Their characteristics are often diagnostic. They are normally alternate on the stem, often folded lengthwise along the centre ("plicate"), and have no stipules. Orchid leaves often have siliceous bodies called stegmata in the vascular bundle sheaths (not present in the Orchidoideae) and are fibrous.

The structure of the leaves corresponds to the specific habitat of the plant. Species that typically bask in sunlight, or grow on sites which can be occasionally very dry, have thick, leathery leaves and the laminae are covered by a waxy cuticle to retain their necessary water supply. Shade-loving species, on the other hand, have long, thin leaves.

The leaves of most orchids are perennial, that is, they live for several years, while others, especially those with plicate leaves as in *Catasetum*, shed them annually and develop new leaves together with new pseudobulbs.

The leaves of some orchids are considered ornamental. The leaves of the *Macodes sanderiana*, a semiterrestrial or rock-hugging ("lithophyte") orchid, show a sparkling silver and gold veining on a light green background. The cordate leaves of *Psychopsis limminghei* are light brownish-green with maroon-puce markings, created by flower pigments. The attractive mottle of the leaves of lady's slippers from tropical and subtropical Asia (*Paphiopedilum*), is caused by uneven distribution of chlorophyll. Also, *Phalaenopsis schilleriana* is a pastel pink orchid with leaves spotted dark green and light green. The jewel orchid (*Ludisia discolor*) is grown more for its colorful leaves than its white flowers.

Some orchids, as *Dendrophylax lindenii* (ghost orchid), *Aphyllorchis* and *Taeniophyllum* depend on their green roots for photosynthesis and lack normally developed leaves, as do all of the heterotrophic species.

Orchids of the genus *Corallorhiza* (coralroot orchids) lack leaves altogether and instead wrap their roots around the roots of mature trees and use specialized fungi to harvest sugars.

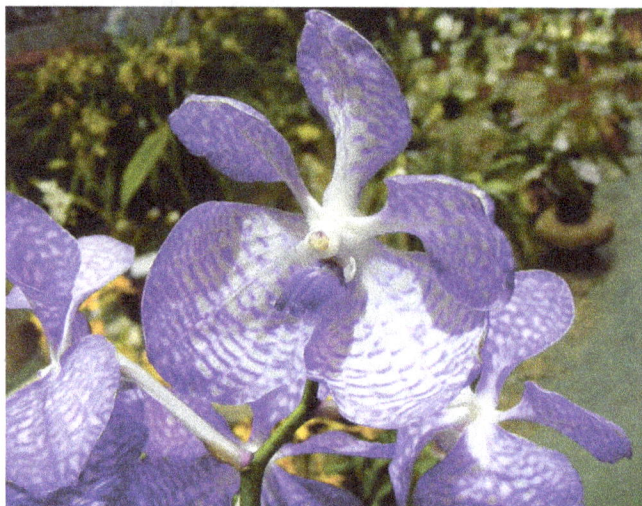
Vanda cultivar

Flowers

The Orchidaceae are well known for the many structural variations in their flowers.

Some orchids have single flowers, but most have a racemose inflorescence, sometimes with a large number of flowers. The flowering stem can be basal, that is, produced from the base of the tuber, like in *Cymbidium*, apical, meaning it grows from the apex of the main stem, like in *Cattleya*, or axillary, from the leaf axil, as in *Vanda*.

As an apomorphy of the clade, orchid flowers are primitively zygomorphic (bilaterally symmetrical), although in some genera like *Mormodes*, *Ludisia*, and *Macodes*, this kind of symmetry may be difficult to notice.

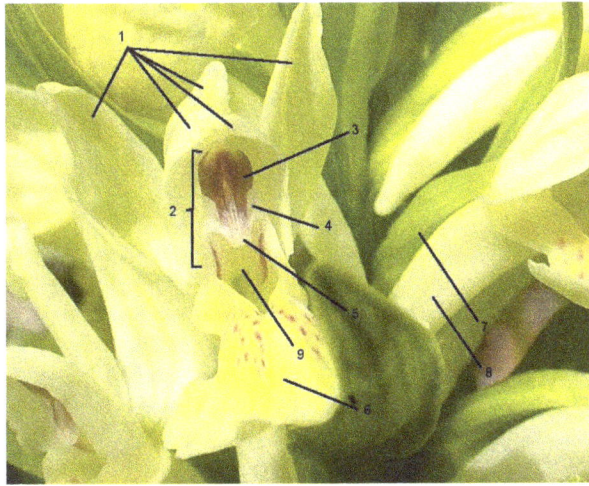

Dactylorhiza sambucina, Orchidoideae for reference

The orchid flower, like most flowers of monocots, has two whorls of sterile elements. The outer whorl has three sepals and the inner whorl has three petals. The sepals are usually very similar to the petals (thus called tepals, 1), but may be completely distinct.

The medial petal, called the labellum or lip (6), which is always modified and enlarged, is actually the upper medial petal; however, as the flower develops, the inferior ovary (7) or the pedicel usually rotates 180°, so that the labellum arrives at the lower part of the flower, thus becoming suitable to form a platform for pollinators. This characteristic, called resupination, occurs primitively in the family and is considered apomorphic, a derived characteristic all Orchidaceae share. The torsion of the ovary is very evident from the longitudinal section shown. Some orchids have secondarily lost this resupination, e.g. *Zygopetalum* and *Epidendrum se-cundum*.

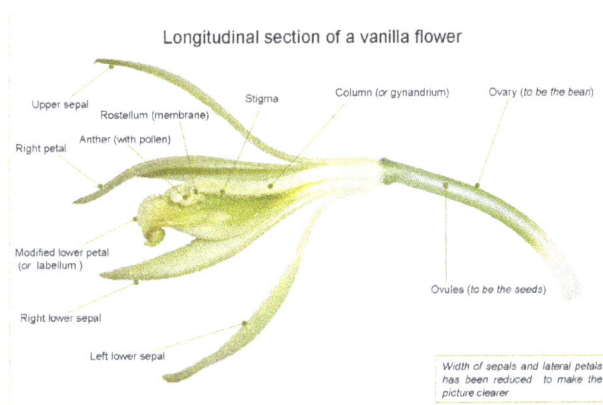

Longitudinal section of a flower of *Vanilla planifolia*

The normal form of the sepals can be found in *Cattleya*, where they form a triangle. In *Paphiope-dilum* (Venus slippers), the lower two sepals are fused into a synsepal, while the lip has taken the form of a slipper. In *Masdevallia*, all the sepals are fused.

Orchid flowers with abnormal numbers of petals or lips are called peloric. Peloria is a genetic trait, but its expression is environmentally influenced and may appear random.

Laeliocattleya cultivar shows the normal form of petals.

Orchid flowers primitively had three stamens, but this situation is now limited to the genus *Neu-wiedia*. *Apostasia* and the Cypripedioideae have two stamens, the central one being sterile and reduced to a staminode. All of the other orchids, the clade called *Monandria*, retain only the central stamen, the others being reduced to staminodes (4). The filaments of the stamens are always adnate (fused) to the style to form cylindrical structure called the gynostemium or column (2). In the primitive Apostasioideae, this fusion is only partial; in the Vanilloideae, it is more deep; in Orchidoideae and Epidendroideae, it is total. The stigma (9) is very asymmetrical, as all of its lobes are bent towards the centre of the flower and lie on the bottom of the column.

Pollen is released as single grains, like in most other plants, in the Apostasioideae, Cypripedioideae, and Vanilloideae. In the other subfamilies, which comprise the great majority of orchids, the anther (3) carries two pollinia.

A pollinium is a waxy mass of pollen grains held together by the glue-like alkaloid viscin, containing both cellulosic strands and mucopolysaccharides. Each pollinium is connected to a filament which can take the form of a caudicle, as in *Dactylorhiza* or *Habenaria*, or a *stipe*, as in *Vanda*. Caudicles or stipes hold the pollinia to the viscidium, a sticky pad which sticks the pollinia to the body of pollinators.

At the upper edge of the stigma of single-anthered orchids, in front of the anther cap, is the rostellum (5), a slender extension involved in the complex pollination mechanism.

As mentioned, the ovary is always inferior (located behind the flower). It is three-carpelate and one or, more rarely, three-partitioned, with parietal placentation (axile in the Apostasioideae).

In 2011, *Bulbophyllum nocturnum* was discovered to flower nocturnally.

Pollination

The complex mechanisms which orchids have evolved to achieve cross-pollination were investigated by Charles Darwin and described in *Fertilisation of Orchids* (1862). Orchids have developed highly specialized pollination systems, thus the chances of being pollinated are often scarce, so orchid flowers usually remain receptive for very long periods, rendering unpollinated flowers long-lasting in cultivation. Most orchids deliver pollen in a single mass. Each time pollination succeeds, thousands of ovules can be fertilized.

Pollinators are often visually attracted by the shape and colours of the labellum. However, some *Bulbophyllum* species attract male fruit flies (*Bactrocera* spp.) solely via a floral chemical which simultaneously acts as a floral reward (e.g. methyl eugenol, raspberry ketone, or zingerone) to perform pollination. The flowers may produce attractive odours. Although absent in most species, nectar may be produced in a spur of the labellum (8 in the illustration above), or on the point of the sepals, or in the septa of the ovary, the most typical position amongst the Asparagales.

In orchids that produce pollinia, pollination happens as some variant of the following sequence: when the pollinator enters into the flower, it touches a viscidium, which promptly sticks to its body, generally on the head or abdomen. While leaving the flower, it pulls the pollinium out of the anther, as it is connected to the viscidium by the caudicle or stipe. The caudicle then bends and the pollinium is moved forwards and downwards. When the pollinator enters another flower of the same species, the pollinium has taken such position that it will stick to the stigma of the second flower, just below the rostellum, pollinating it. The possessors of orchids may be able to reproduce the process with a pencil, small paintbrush, or other similar device.

Ophrys apifera is about to self-pollinate

Some orchids mainly or totally rely on self-pollination, especially in colder regions where pollinators are particularly rare. The caudicles may dry up if the flower has not been visited by any pollinator, and the pollinia then fall directly on the stigma. Otherwise, the anther may rotate and then enter the stigma cavity of the flower (as in *Holcoglossum amesianum*).

The slipper orchid *Paphiopedilum parishii* reproduces by self-fertilization. This occurs when the anther changes from a solid to a liquid state and directly contacts the stigma surface without the aid of any pollinating agent or floral assembly.

The labellum of the Cypripedioideae is poke bonnet-shaped, and has the function of trapping visiting insects. The only exit leads to the anthers that deposit pollen on the visitor.

In some extremely specialized orchids, such as the Eurasian genus *Ophrys*, the labellum is adapted to have a colour, shape, and odour which attracts male insects via mimicry of a receptive female. Pollination happens as the insect attempts to mate with flowers.

Many neotropical orchids are pollinated by male orchid bees, which visit the flowers to gather volatile chemicals they require to synthesize pheromonal attractants. Males of such species as *Euglossa imperialis* or *Eulaema meriana* have been observed to leave their territories periodically to forage for aromatic compounds, such as cineole, to synthesize pheromone for attracting and mating with females. Each type of orchid places the pollinia on a different body part of a different species of bee, so as to enforce proper cross-pollination.

A rare achlorophyllous saprophytic orchid growing entirely underground in Australia, *Rhizanthella slateri*, is never exposed to light, and depends on ants and other terrestrial insects to pollinate it.

Catasetum, a genus discussed briefly by Darwin, actually launches its viscid pollinia with explosive force when an insect touches a seta, knocking the pollinator off the flower.

After pollination, the sepals and petals fade and wilt, but they usually remain attached to the ovary.

Asexual Reproduction

Some species, such as *Phalaenopsis*, *Dendrobium*, and *Vanda*, produce offshoots or plantlets formed from one of the nodes along the stem, through the accumulation of growth hormones at that point. These shoots are known as *keiki*.

Fruits and Seeds

Cross-sections of orchid capsules showing the longitudinal slits

The ovary typically develops into a capsule that is dehiscent by three or six longitudinal slits, while remaining closed at both ends.

The seeds are generally almost microscopic and very numerous, in some species over a million per capsule. After ripening, they blow off like dust particles or spores. They lack endosperm and must enter symbiotic relationships with various mycorrhizal basidiomyceteous fungi that provide them the necessary nutrients to germinate, so all orchid species are mycoheterotrophic during germination and reliant upon fungi to complete their lifecycles.

Closeup of a *Phalaenopsis* blossom

As the chance for a seed to meet a suitable fungus is very small, only a minute fraction of all the seeds released grow into adult plants. In cultivation, germination typically takes weeks.

Horticultural techniques have been devised for germinating orchid seeds on an artificial nutrient medium, eliminating the requirement of the fungus for germination and greatly aiding the propagation of ornamental orchids. The usual medium for the sowing of orchids in artificial conditions is agar agar gel combined with a carbohydrate energy source. The carbohydrate source can be combinations of discrete sugars or can be derived from other sources such as banana, pineapple, peach, or even tomato puree or coconut water. After the preparation of the agar agar medium, it is poured into test tubes or jars which are then autoclaved (or cooked in a pressure cooker) to sterilize the medium. After cooking, the medium begins to gel as it cools.

Taxonomy

The taxonomy of this family is in constant flux, as new studies continue to clarify the relationships between species and groups of species, allowing more taxa at several ranks to be recognized. The Orchidaceae is currently placed in the order Asparagales by the APG III system of 2009.

Five subfamilies are recognised. The cladogram below was made according to the APG system of 1998. It represents the view that most botanists had held up to that time. It was supported by morphological studies, but never received strong support in molecular phylogenetic studies.

In 2015, a phylogenetic study showed strong statistical support for the following topology of the orchid tree, using 9 kb of plastid and nuclear DNA from 7 genes, a topology that was confirmed by a phylogenomic study in the same year.

Evolution

A study in the scientific journal *Nature* has hypothesised that the origin of orchids goes back much longer than originally expected. An extinct species of stingless bee, *Proplebeia dominicana*, was found trapped in Miocene amber from about 15-20 million years ago. The bee was carrying pollen of a previously unknown orchid taxon, *Meliorchis caribea*, on its wings. This find is the first evidence of fossilised orchids to date and shows insects were active pollinators of orchids then. This extinct orchid, *M. caribea*, has been placed within the extant tribe Cranichideae, subtribe Goodyerinae (subfamily Orchidoideae). An even older orchid species, *Succinanthera baltica*, was described from the Eocene Baltic amber by Poinar & Rasmussen (2017).

Genetic sequencing indicates orchids may have arisen earlier, 76 to 84 million years ago during the Late Cretaceous. According to Mark W. Chase *et al.* (2001), the overall biogeography and phylogenetic patterns of Orchidaceae show they are even older and may go back roughly 100 million years.

Using the molecular clock method, it was possible to determine the age of the major branches of the orchid family. This also confirmed that the subfamily Vanilloideae is a branch at the basal dichotomy of the monandrous orchids, and must have evolved very early in the evolution of the family. Since this subfamily occurs worldwide in tropical and subtropical regions, from tropical America to tropical Asia, New Guinea and West Africa, and the continents began to split about 100 million years ago, significant biotic exchange must have occurred after this split (since the age of *Vanilla* is estimated at 60 to 70 million years).

Etymology

The type genus (i.e. the genus after which the family is named) is *Orchis*. The genus name comes from the Ancient Greek *órkhis*, literally meaning "testicle", because of the shape of the twin tubers in some species of *Orchis*. The term "orchid" was introduced in 1845 by John Lindley in *School Botany*, as a shortened form of *Orchidaceae*.

Distribution

Orchidaceae are cosmopolitan, occurring in almost every habitat apart from glaciers. The world's richest diversity of orchid genera and species is found in the tropics, but they are also found above the Arctic Circle, in southern Patagonia, and two species of *Nematoceras* on Macquarie Island at 54° south.

The following list gives a rough overview of their distribution:

- Oceania: 50 to 70 genera
- North America: 20 to 26 genera
- tropical America: 212 to 250 genera
- tropical Asia: 260 to 300 genera
- tropical Africa: 230 to 270 genera
- Europe and temperate Asia: 40 to 60 genera

Ecology

A majority of orchids are perennial epiphytes, which grow anchored to trees or shrubs in the tropics and subtropics. Species such as *Angraecum sororium* are lithophytes, growing on rocks or very rocky soil. Other orchids (including the majority of temperate Orchidaceae) are terrestrial and can be found in habitat areas such as grasslands or forest.

Some orchids, such as *Neottia* and *Corallorhiza*, lack chlorophyll, so are unable to photosynthesise. Instead, these species obtain energy and nutrients by parasitising soil fungi through the formation of orchid mycorrhizas. The fungi involved include those that form ectomycorrhizas with trees and other woody plants, parasites such as *Armillaria*, and saprotrophs. These orchids are known as myco-heterotrophs, but were formerly (incorrectly) described as saprophytes as it was believed they gained their nutrition by breaking down organic matter. While only a few species are achlorophyllous holoparasites, all orchids are myco-heterotrophic during germination and seedling growth, and even photosynthetic adult plants may continue to obtain carbon from their mycorrhizal fungi.

Uses

As decoration in a flowerpot

A flower of a Blc. Paradise Jewel 'Flame' hybrid orchid plant. Blooms of the *Cattleya* alliance are often used in ladies' corsages.

Perfumery

The scent of orchids is frequently analysed by perfumers (using headspace technology and gas-liquid chromatography/mass spectrometry) to identify potential fragrance chemicals.

Horticulture

The other important use of orchids is their cultivation for the enjoyment of the flowers. Most cultivated orchids are tropical or subtropical, but quite a few which grow in colder climates can be found on the market. Temperate species available at nurseries include *Ophrys apifera* (bee orchid), *Gymnadenia conopsea* (fragrant orchid), *Anacamptis pyramidalis* (pyramidal orchid) and *Dactylorhiza fuchsii* (common spotted orchid).

Orchids of all types have also often been sought by collectors of both species and hybrids. Many hundreds of societies and clubs worldwide have been established. These can be small, local clubs, or larger, national organisations such as the American Orchid Society. Both serve to encourage cultivation and collection of orchids, but some go further by concentrating on conservation or research.

The term "botanical orchid" loosely denotes those small-flowered, tropical orchids belonging to several genera that do not fit into the "florist" orchid category. A few of these genera contain enormous numbers of species. Some, such as *Pleurothallis* and *Bulbophyllum*, contain approximately 1700 and 2000 species, respectively, and are often extremely vegetatively diverse. The primary use of the term is among orchid hobbyists wishing to describe unusual species they grow, though it is also used to distinguish naturally occurring orchid species from horticulturally created hybrids.

Use as Food

Vanilla fruits drying

The dried seed pods of one orchid genus, *Vanilla* (especially *Vanilla planifolia*), are commercially important as a flavouring in baking, for perfume manufacture and aromatherapy.

The underground tubers of terrestrial orchids [mainly *Orchis mascula* (early purple orchid)] are ground to a powder and used for cooking, such as in the hot beverage *salep* or in the Turkish frozen treat *dondurma*. The name *salep* has been claimed to come from the Arabic expression ḥasyu al-tha 'lab, "fox testicles", but it appears more likely the name comes directly from the Arabic name saḥlab. The similarity in appearance to testes naturally accounts for *salep* being considered an aphrodisiac.

The dried leaves of *Jumellea fragrans* are used to flavour rum on Reunion Island.

Some saprophytic orchid species of the group *Gastrodia* produce potato-like tubers and were consumed as food by native peoples in Australia and can be successfully cultivated, notably *Gastrodia sesamoides*. Wild stands of these plants can still be found in the same areas as early aboriginal settlements, such as Ku-ring-gai Chase National Park in Australia. Aboriginal peoples located the plants in habitat by observing where bandicoots had scratched in search of the tubers after detecting the plants underground by scent.

Traditional Medicinal Uses

Orchids have been used in traditional medicine in an effort to treat many diseases and ailments. They have been used as a source of herbal remedies in China since 2800 BC. *Gastrodia elata* is one of the three orchids listed in the earliest known Chinese Materia Medica (*Shennon bencaojing*) (c. 100 AD). Theophrastus mentions orchids in his *Enquiry into Plants* (372–286 BC).

Cultural Symbolism

Orchids have many associations with symbolic values. For example, the orchid is the City Flower of Shaoxing, China. *Cattleya mossiae* is the national Venezuelan flower, while *Cattleya trianae* is the national flower of Colombia. *Vanda* 'Miss Joaquim' is the national flower of Singapore, *Guarianthe skinneri* is the national flower of Costa Rica and *Rhyncholaelia digbyana* is the national flower of Honduras. *Prosthechea cochleata* is the national flower of Belize, where it is known as the *black orchid*. *Lycaste skinneri* has a white variety (alba) which is the national flower of Guatemala, commonly known as *Monja Blanca* (White Nun). Panama's national flower is the *Holy Ghost orchid* (*Peristeria elata*), or 'the flor del Espiritu Santo'.

Orchids native to the Mediterranean are depicted on the *Ara Pacis* in Rome, until now the only known instance of orchids in ancient art, and the earliest in European art.

References

- Crane, P.R.; Herendeen, P. & Friis, E.M. (2004), "Fossils and plant phylogeny", American Journal of Botany, 91 (10): 1683–99, PMID 21652317, doi:10.3732/ajb.91.10.1683, retrieved 2011-01-28

- Kenrick, P. & Crane, P.R. (1997), The Origin and Early Diversification of Land Plants: A Cladistic Study, Washington, D.C.: Smithsonian Institution Press, ISBN 978-1-56098-730-7

- Niklas, K.J.; Kutschera, U. (2010), "The evolution of the land plant life cycle", New Phytologist, 185 (1): 27–41, PMID 19863728, doi:10.1111/j.1469-8137.2009.03054.x

- Rutishauser, R. (1999), "Polymerous Leaf Whorls in Vascular Plants: Developmental Morphology and Fuzziness of Organ Identities", International Journal of Plant Sciences, 160 (6): 81–103, PMID 10572024, doi:10.1086/314221

- Novíkov & Barabaš-Krasni (2015). "Modern plant systematics". Liga-Pres: 685. ISBN 978-966-397-276-3. doi:10.13140/RG.2.1.4745.6164

- "Algae Herbarium". National Museum of Natural History, Department of Botany. 2008. Archived from the original on 1 December 2008. Retrieved 19 December 2008

- Bhattacharya, D.; Medlin, L. (1998). "Algal Phylogeny and the Origin of Land Plants" (PDF). Plant Physiology. 116 (1): 9–15. doi:10.1104/pp.116.1.9

- Tarakhovskaya, E. R.; Maslov, Yu. I.; Shishova, M. F. (April 2007). "Phytohormones in algae". Russian Journal of Plant Physiology. 54 (2): 163–170. doi:10.1134/s1021443707020021

- Anderson, Anderson & Cleal (2007). "Brief history of the gymnosperms: classification, biodiversity, phytogeography and ecology". Strelitzia. SANBI. 20: 280. ISBN 978-1-919976-39-6

- National Recovery Plan for the MacDonnell Ranges Cycad Macrozamia macdonnellii (PDF), Department of Natural Resources, Environment, The Arts and Sport, Northern Territory, retrieved 16 July 2015

- Archibald JM; Keeling PJ (November 2002). "Recycled plastids: a 'green movement' in eukaryotic evolution". Trends in Genetics. 18 (11): 577–584. PMID 12414188. doi:10.1016/S0168-9525(02)02777-4

- "ALGAL RESPONSE TO NUTRIENT ENRICHMENT IN FORESTED OLIGOTROPHIC STREAM". Journal of Phycology. 44: 564–572. doi:10.1111/j.1529-8817.2008.00503.x

- Taylor, T.N.; Taylor, E.L. & Krings, M. (2009), Paleobotany, The Biology and Evolution of Fossil Plants (2nd ed.), Amsterdam; Boston: Academic Press, ISBN 978-0-12-373972-8 , pp. 508ff

- Round, FE (1981). "Chapter 8, Dispersal, continuity and phytogeography". The ecology of algae. pp. 357–361. Retrieved 6 February 2015

- Wellman, C.H.; Osterloff, P.L.; Mohiuddin, U. (2003). "Fragments of the earliest land plants". Nature. 425 (6955): 282–285. Bibcode:2003Natur.425..282W. PMID 13679913. doi:10.1038/nature01884

- Crane, P. R.; Herendeen, P. & Friis, E. M. (2004). "Fossils and plant phylogeny". American Journal of Botany. 91 (10): 1683–99. PMID 21652317. doi:10.3732/ajb.91.10.1683

- Silva PC, Basson PW and Moe RL (1996) Catalogue of the Benthic Marine Algae of the Indian Ocean page 2, University of California Press. ISBN 978-0-520-91581-7

- Otto SP (2009). "The evolutionary enigma of sex". Am. Nat. 174 Suppl 1: S1–S14. PMID 19441962. doi:10.1086/599084

- Jeffrey D. Palmer; Douglas E. Soltis; Mark W. Chase (2004). "The plant tree of life: an overview and some points of view". American Journal of Botany. 91 (10): 1437–1445. PMID 21652302. doi:10.3732/ajb.91.10.1437

- Taylor, T.N.; Taylor, E.L. & Krings, M. (2009), Paleobotany, The Biology and Evolution of Fossil Plants (2nd ed.), Amsterdam; Boston: Academic Press, ISBN 978-0-12-373972-8 , p. 1027

- "Secondary Products of Brown Algae". Algae Research. Smithsonian National Museum of Natural History. Retrieved 29 December 2008

- Cronberg N, Natcheva R, Hedlund K (2006). "Microarthropods Mediate Sperm Transfer in Mosses". Science. 313 (5791): 1255. PMID 16946062. doi:10.1126/science.1128707

- Soltis, D. E.; Soltis, P. S. (2004). "Amborella not a "basal angiosperm"? Not so fast". American Journal of Botany. 91 (6): 997–1001. PMID 21653455. doi:10.3732/ajb.91.6.997

Plant Physiology: An Integrated Study

Plant physiology studies the functioning of a plant. The processes studied under this subject are plant nutrition, photoperiodism, plant hormone, plant pathology and photomorphogenesis. The section on plant physiology offers an insightful focus, keeping in mind the complex subject matter.

Plant Physiology

A germination rate experiment

Plant physiology is a subdiscipline of botany concerned with the functioning, or physiology, of plants. Closely related fields include plant morphology (structure of plants), plant ecology (interactions with the environment), phytochemistry (biochemistry of plants), cell biology, genetics, biophysics and molecular biology.

Fundamental processes such as photosynthesis, respiration, plant nutrition, plant hormone functions, tropisms, nastic movements, photoperiodism, photomorphogenesis, circadian rhythms, environmental stress physiology, seed germination, dormancy and stomata function and transpiration, both parts of plant water relations, are studied by plant physiologists.

Aims

The field of plant physiology includes the study of all the internal activities of plants—those chemical and physical processes associated with life as they occur in plants. This includes study at many levels of scale of size and time. At the smallest scale are molecular interactions of photosynthesis and internal diffusion of water, minerals, and nutrients. At the largest scale are the processes of plant development, seasonality, dormancy, and reproductive control. Major subdisciplines of plant physiology include phytochemistry (the study of the biochemistry of plants) and phytopathology (the study of disease in plants). The scope of plant physiology as a discipline may be divided into several major areas of research.

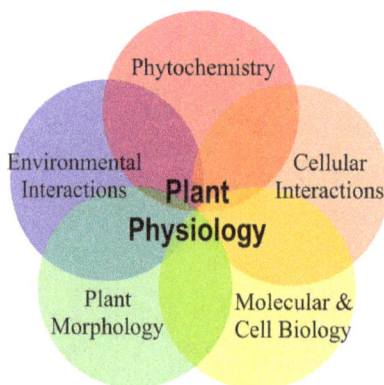

Five key areas of study within plant physiology

First, the study of phytochemistry (plant chemistry) is included within the domain of plant physiology. To function and survive, plants produce a wide array of chemical compounds not found in other organisms. Photosynthesis requires a large array of pigments, enzymes, and other compounds to function. Because they cannot move, plants must also defend themselves chemically from herbivores, pathogens and competition from other plants. They do this by producing toxins and foul-tasting or smelling chemicals. Other compounds defend plants against disease, permit survival during drought, and prepare plants for dormancy, while other compounds are used to attract pollinators or herbivores to spread ripe seeds.

Secondly, plant physiology includes the study of biological and chemical processes of individual plant cells. Plant cells have a number of features that distinguish them from cells of animals, and which lead to major differences in the way that plant life behaves and responds differently from animal life. For example, plant cells have a cell wall which restricts the shape of plant cells and thereby limits the flexibility and mobility of plants. Plant cells also contain chlorophyll, a chemical compound that interacts with light in a way that enables plants to manufacture their own nutrients rather than consuming other living things as animals do.

Thirdly, plant physiology deals with interactions between cells, tissues, and organs within a plant. Different cells and tissues are physically and chemically specialized to perform different functions. Roots and rhizoids function to anchor the plant and acquire minerals in the soil. Leaves catch light in order to manufacture nutrients. For both of these organs to remain living, minerals that the roots acquire must be transported to the leaves, and the nutrients manufactured in the leaves must be transported to the roots. Plants have developed a number of ways to achieve this transport, such as vascular tissue, and the functioning of the various modes of transport is studied by plant physiologists.

Fourthly, plant physiologists study the ways that plants control or regulate internal functions. Like animals, plants produce chemicals called hormones which are produced in one part of the plant to signal cells in another part of the plant to respond. Many flowering plants bloom at the appropriate time because of light-sensitive compounds that respond to the length of the night, a phenomenon known as photoperiodism. The ripening of fruit and loss of leaves in the winter are controlled in part by the production of the gas ethylene by the plant.

Finally, plant physiology includes the study of plant response to environmental conditions and their variation, a field known as environmental physiology. Stress from water loss, changes in air chemistry, or crowding by other plants can lead to changes in the way a plant functions. These changes may be affected by genetic, chemical, and physical factors.

Biochemistry of Plants

Latex being collected from a tapped rubber tree

The chemical elements of which plants are constructed—principally carbon, oxygen, hydrogen, nitrogen, phosphorus, sulfur, etc.—are the same as for all other life forms animals, fungi, bacteria and even viruses. Only the details of the molecules into which they are assembled differs.

Despite this underlying similarity, plants produce a vast array of chemical compounds with unique properties which they use to cope with their environment. Pigments are used by plants to absorb or detect light, and are extracted by humans for use in dyes. Other plant products may be used for the manufacture of commercially important rubber or biofuel. Perhaps the most celebrated compounds from plants are those with pharmacological activity, such as salicylic acid from which aspirin is made, morphine, and digoxin. Drug companies spend billions of dollars each year researching plant compounds for potential medicinal benefits.

Constituent Elements

Plants require some nutrients, such as carbon and nitrogen, in large quantities to survive. Such nutrients are termed macronutrients, where the prefix *macro-* (large) refers to the quantity needed, not the size of the nutrient particles themselves. Other nutrients, called micronutrients, are required only in trace amounts for plants to remain healthy. Such micronutrients are usually absorbed as ions dissolved in water taken from the soil, though carnivorous plants acquire some of their micronutrients from captured prey.

The following tables list element nutrients essential to plants. Uses within plants are generalized.

Macronutrients – necessary in large quantities		
Element	Form of uptake	Notes
Nitrogen	NO_3^-, NH_4^+	Nucleic acids, proteins, hormones, etc.
Oxygen	O_2 H_2O	Cellulose, starch, other organic compounds
Carbon	CO_2	Cellulose, starch, other organic compounds
Hydrogen	H_2O	Cellulose, starch, other organic compounds
Potassium	K^+	Cofactor in protein synthesis, water balance, etc.
Calcium	Ca^{2+}	Membrane synthesis and stabilization
Magnesium	Mg^{2+}	Element essential for chlorophyll
Phosphorus	$H_2PO_4^-$	Nucleic acids, phospholipids, ATP
Sulfur	SO_4^{2-}	Constituent of proteins

Micronutrients – necessary in small quantities		
Element	Form of uptake	Notes
Chlorine	Cl^-	Photosystem II and stomata function
Iron	Fe^{2+}, Fe^{3+}	Chlorophyll formation
Boron	HBO_3	Crosslinking pectin
Manganese	Mn^{2+}	Activity of some enzymes
Zinc	Zn^{2+}	Involved in the synthesis of enzymes and chlorophyll
Copper	Cu^+	Enzymes for lignin synthesis
Molybdenum	MoO_4^{2-}	Nitrogen fixation, reduction of nitrates
Nickel	Ni^{2+}	Enzymatic cofactor in the metabolism of nitrogen compounds

Pigments

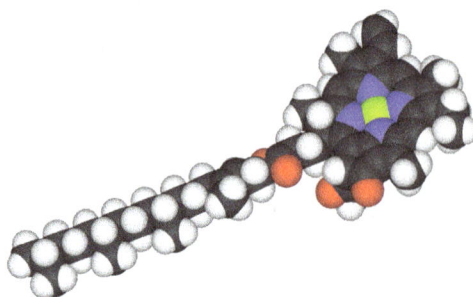

Space-filling model of the chlorophyll molecule

Anthocyanin gives these pansies their dark purple pigmentation

Among the most important molecules for plant function are the pigments. Plant pigments include a variety of different kinds of molecules, including porphyrins, carotenoids, and anthocyanins. All biological pigments selectively absorb certain wavelengths of light while reflecting others. The light that is absorbed may be used by the plant to power chemical reactions, while the reflected wavelengths of light determine the color the pigment appears to the eye.

Chlorophyll is the primary pigment in plants; it is a porphyrin that absorbs red and blue wavelengths of light while reflecting green. It is the presence and relative abundance of chlorophyll that gives plants their green color. All land plants and green algae possess two forms of this pigment: chlorophyll *a* and chlorophyll *b*. Kelps, diatoms, and other photosynthetic heterokonts contain chlorophyll *c* instead of *b*, red algae possess chlorophyll *a* and " d". All chlorophylls serve as the primary means plants use to intercept light to fuel photosynthesis.

Carotenoids are red, orange, or yellow tetraterpenoids. They function as accessory pigments in plants, helping to fuel photosynthesis by gathering wavelengths of light not readily absorbed by chlorophyll. The most familiar carotenoids are carotene (an orange pigment found in carrots), lutein (a yellow pigment found in fruits and vegetables), and lycopene (the red pigment responsible for the color of tomatoes). Carotenoids have been shown to act as antioxidants and to promote healthy eyesight in humans.

Anthocyanins (literally "flower blue") are water-soluble flavonoid pigments that appear red to blue, according to pH. They occur in all tissues of higher plants, providing color in leaves, stems, roots, flowers, and fruits, though not always in sufficient quantities to be noticeable. Anthocyanins are most visible in the petals of flowers, where they may make up as much as 30% of the dry weight of the tissue. They are also responsible for the purple color seen on the underside of tropical shade plants such as *Tradescantia zebrina*. In these plants, the anthocyanin catches light that has passed through the leaf and reflects it back towards regions bearing chlorophyll, in order to maximize the use of available light

Betalains are red or yellow pigments. Like anthocyanins they are water-soluble, but unlike anthocyanins they are indole-derived compounds synthesized from tyrosine. This class of pigments is found only in the Caryophyllales (including cactus and amaranth), and never co-occur in plants with anthocyanins. Betalains are responsible for the deep red color of beets, and are used commercially as food-coloring agents. Plant physiologists are uncertain of the function that betalains have in plants which possess them, but there is some preliminary evidence that they may have fungicidal properties.

Signals and Regulators

A mutation that stops *Arabidopsis thaliana* responding to auxin causes abnormal growth (right)

Plants produce hormones and other growth regulators which act to signal a physiological response in their tissues. They also produce compounds such as phytochrome that are sensitive to light and which serve to trigger growth or development in response to environmental signals.

Plant Hormones

Plant hormones, known as plant growth regulators (PGRs) or phytohormones, are chemicals that regulate a plant's growth. According to a standard animal definition, hormones are signal molecules produced at specific locations, that occur in very low concentrations, and cause altered processes in target cells at other locations. Unlike animals, plants lack specific hormone-producing tissues or organs. Plant hormones are often not transported to other parts of the plant and production is not limited to specific locations.

Plant hormones are chemicals that in small amounts promote and influence the growth, development and differentiation of cells and tissues. Hormones are vital to plant growth; affecting processes in plants from flowering to seed development, dormancy, and germination. They regulate which tissues grow upwards and which grow downwards, leaf formation and stem growth, fruit development and ripening, as well as leaf abscission and even plant death.

The most important plant hormones are abscissic acid (ABA), auxins, ethylene, gibberellins, and cytokinins, though there are many other substances that serve to regulate plant physiology.

Photomorphogenesis

While most people know that light is important for photosynthesis in plants, few realize that plant sensitivity to light plays a role in the control of plant structural development (morphogenesis).

The use of light to control structural development is called photomorphogenesis, and is dependent upon the presence of specialized photoreceptors, which are chemical pigments capable of absorbing specific wavelengths of light.

Plants use four kinds of photoreceptors: phytochrome, cryptochrome, a UV-B photoreceptor, and protochlorophyllide a. The first two of these, phytochrome and cryptochrome, are photoreceptor proteins, complex molecular structures formed by joining a protein with a light-sensitive pigment. Cryptochrome is also known as the UV-A photoreceptor, because it absorbs ultraviolet light in the long wave "A" region. The UV-B receptor is one or more compounds not yet identified with certainty, though some evidence suggests carotene or riboflavin as candidates. Protochlorophyllide a, as its name suggests, is a chemical precursor of chlorophyll.

The most studied of the photoreceptors in plants is phytochrome. It is sensitive to light in the red and far-red region of the visible spectrum. Many flowering plants use it to regulate the time of flowering based on the length of day and night (photoperiodism) and to set circadian rhythms. It also regulates other responses including the germination of seeds, elongation of seedlings, the size, shape and number of leaves, the synthesis of chlorophyll, and the straightening of the epicotyl or hypocotyl hook of dicot seedlings.

Photoperiodism

The poinsettia is a short-day plant, requiring two months of long nights prior to blooming.

Many flowering plants use the pigment phytochrome to sense seasonal changes in day length, which they take as signals to flower. This sensitivity to day length is termed photoperiodism. Broadly speaking, flowering plants can be classified as long day plants, short day plants, or day neutral plants, depending on their particular response to changes in day length. Long day plants require a certain minimum length of daylight to starts flowering, so these plants flower in the spring or summer. Conversely, short day plants flower when the length of daylight falls below a certain critical level. Day neutral plants do not initiate flowering based on photoperiodism, though some may use temperature sensitivity (vernalization) instead.

Although a short day plant cannot flower during the long days of summer, it is not actually the period of light exposure that limits flowering. Rather, a short day plant requires a minimal length of uninterrupted darkness in each 24-hour period (a short daylength) before floral development can begin. It has been determined experimentally that a short day plant (long night) does not flower if a flash of phytochrome activating light is used on the plant during the night.

Plants make use of the phytochrome system to sense day length or photoperiod. This fact is utilized by florists and greenhouse gardeners to control and even induce flowering out of season, such as the *Poinsettia*.

Environmental Physiology

Phototropism in *Arabidopsis thaliana* is regulated by blue to UV light

Paradoxically, the subdiscipline of environmental physiology is on the one hand a recent field of study in plant ecology and on the other hand one of the oldest. Environmental physiology is the preferred name of the subdiscipline among plant physiologists, but it goes by a number of other names in the applied sciences. It is roughly synonymous with ecophysiology, crop ecology, horticulture and agronomy. The particular name applied to the subdiscipline is specific to the viewpoint and goals of research. Whatever name is applied, it deals with the ways in which plants respond to their environment and so overlaps with the field of ecology.

Environmental physiologists examine plant response to physical factors such as radiation (including light and ultraviolet radiation), temperature, fire, and wind. Of particular importance are water relations (which can be measured with the Pressure bomb) and the stress of drought or inundation, exchange of gases with the atmosphere, as well as the cycling of nutrients such as nitrogen and carbon.

Environmental physiologists also examine plant response to biological factors. This includes not only negative interactions, such as competition, herbivory, disease and parasitism, but also positive interactions, such as mutualism and pollination.

Tropisms and Nastic Movements

Plants may respond both to directional and non-directional stimuli. A response to a directional stimulus, such as gravity or sunlight, is called a tropism. A response to a nondirectional stimulus, such as temperature or humidity, is a nastic movement.

Tropisms in plants are the result of differential cell growth, in which the cells on one side of the plant elongates more than those on the other side, causing the part to bend toward the side with less growth. Among the common tropisms seen in plants is phototropism, the bending of the plant toward a source of light. Phototropism allows the plant to maximize light exposure in plants which require additional light for photosynthesis, or to minimize it in plants subjected to intense light and heat. Geotropism allows the roots of a plant to determine the direction of gravity and grow downwards. Tropisms generally result from an interaction between the environment and production of one or more plant hormones.

Nastic movements results from differential cell growth (e.g. epinasty and hiponasty), or from changes in turgor pressure within plant tissues (e.g., nyctinasty), which may occur rapidly. A familiar example is thigmonasty (response to touch) in the Venus fly trap, a carnivorous plant. The traps consist of modified leaf blades which bear sensitive trigger hairs. When the hairs are touched by an insect or other animal, the leaf folds shut. This mechanism allows the plant to trap and digest small insects for additional nutrients. Although the trap is rapidly shut by changes in internal cell pressures, the leaf must grow slowly to reset for a second opportunity to trap insects.

Plant Disease

Powdery mildew on crop leaves

Economically, one of the most important areas of research in environmental physiology is that of phytopathology, the study of diseases in plants and the manner in which plants resist or cope with infection. Plant are susceptible to the same kinds of disease organisms as animals, including viruses, bacteria, and fungi, as well as physical invasion by insects and roundworms.

Because the biology of plants differs with animals, their symptoms and responses are quite different. In some cases, a plant can simply shed infected leaves or flowers to prevent the spread of disease, in a process called abscission. Most animals do not have this option as a means of controlling disease. Plant diseases organisms themselves also differ from those causing disease in animals because plants cannot usually spread infection through casual physical contact. Plant pathogens tend to spread via spores or are carried by animal vectors.

One of the most important advances in the control of plant disease was the discovery of Bordeaux mixture in the nineteenth century. The mixture is the first known fungicide and is a combination of copper sulfate and lime. Application of the mixture served to inhibit the growth of downy mildew that threatened to seriously damage the French wine industry.

History

Early History

Jan Baptist van Helmont

Sir Francis Bacon published one of the first plant physiology experiments in 1627 in the book, *Sylva Sylvarum*. Bacon grew several terrestrial plants, including a rose, in water and concluded that soil was only needed to keep the plant upright. Jan Baptist van Helmont published what is considered the first quantitative experiment in plant physiology in 1648. He grew a willow tree for five years in a pot containing 200 pounds of oven-dry soil. The soil lost just two ounces of dry weight and van Helmont concluded that plants get all their weight from water, not soil. In 1699, John Woodward published experiments on growth of spearmint in different sources of water. He found that plants grew much better in water with soil added than in distilled water.

Stephen Hales is considered the Father of Plant Physiology for the many experiments in the 1727 book; though Julius von Sachs unified the pieces of plant physiology and put them together as a discipline. His *Lehrbuch der Botanik* was the plant physiology bible of its time.

Researchers discovered in the 1800s that plants absorb essential mineral nutrients as inorganic ions in water. In natural conditions, soil acts as a mineral nutrient reservoir but the soil itself is not essential to plant growth. When the mineral nutrients in the soil are dissolved in water, plant roots absorb nutrients readily, soil is no longer required for the plant to thrive. This observation is the basis for hydroponics, the growing of plants in a water solution rather than soil, which has become a standard technique in biological research, teaching lab exercises, crop production and as a hobby.

Current Research

One of the leading journals in the field is *Plant Physiology*, started in 1926. All its back issues are available online for free. Many other journals often carry plant physiology articles, including *Physiologia Plantarum, Journal of Experimental Botany, American Journal of Botany, Annals of Botany, Journal of Plant Nutrition* and *Proceedings of the National Academy of Sciences.*

Economic Applications

Food Production

In horticulture and agriculture along with food science, plant physiology is an important topic relating to fruits, vegetables, and other consumable parts of plants. Topics studied include: *climatic* requirements, fruit drop, nutrition, ripening, fruit set. The production of food crops also hinges on the study of plant physiology covering such topics as optimal planting and harvesting times and post harvest storage of plant products for human consumption and the production of secondary products like drugs and cosmetics.

Plant Nutrition

Plant nutrition is the study of the chemical elements and compounds necessary for plant growth, plant metabolism and their external supply. In 1972, Emanuel Epstein defined two criteria for an element to be essential for plant growth:

1. in its absence the plant is unable to complete a normal life cycle.

2. or that the element is part of some essential plant constituent or metabolite.

This is in accordance with Justus von Liebig's law of the minimum. The essential plant nutrients include carbon, oxygen and hydrogen which are absorbed from the air, whereas other nutrients including nitrogen are typically obtained from the soil (exceptions include some parasitic or carnivorous plants).

Plants must obtain the following mineral nutrients from their growing medium:

*   the macronutrients: nitrogen (N), phosphorus (P), potassium (K), calcium (Ca), sulfur (S), magnesium (Mg), sodium (Na)

*   the micronutrients (or trace minerals): boron (B), chlorine (Cl), manganese (Mn), iron (Fe), zinc (Zn), copper (Cu), molybdenum (Mo), nickel (Ni). and cobalt (Co)

The macronutrients are consumed in larger quantities; hydrogen, oxygen, nitrogen and carbon contribute to over 95% of a plants' entire biomass on a dry matter weight basis. Micronutrients are present in plant tissue in quantities measured in parts per million, ranging from 0.1 to 200 ppm, or less than 0.02% dry weight.

Most soil conditions across the world can provide plants adapted to that climate and soil with sufficient nutrition for a complete life cycle, without the addition of nutrients as fertilizer. However, if the soil is cropped it is necessary to artificially modify soil fertility through the addition of fertilizer to promote vigorous growth and increase or sustain yield. This is done because, even with adequate water and light, nutrient deficiency can limit growth and crop yield.

Farmer spreading decomposing manure to improve soil fertility and plant nutrition

Processes

Plants take up essential elements from the soil through their roots and from the air (mainly consisting of nitrogen and oxygen) through their leaves. Nutrient uptake in the soil is achieved by cation exchange, wherein root hairs pump hydrogen ions (H^+) into the soil through proton pumps. These hydrogen ions displace cations attached to negatively charged soil particles so that the cations are available for uptake by the root. In the leaves, stomata open to take in carbon dioxide and expel oxygen. The carbon dioxide molecules are used as the carbon source in photosynthesis.

The root, especially the root hair, is the essential organ for the uptake of nutrients. The structure and architecture of the root can alter the rate of nutrient uptake. Nutrient ions are transported to the center of the root, the stele, in order for the nutrients to reach the conducting tissues, xylem and phloem. The Casparian strip, a cell wall outside the stele but within the root, prevents passive flow of water and nutrients, helping to regulate the uptake of nutrients and water. Xylem moves water and mineral ions within the plant and phloem accounts for organic molecule transportation. Water potential plays a key role in a plant's nutrient uptake. If the water potential is more negative within the plant than the surrounding soils, the nutrients will move from the region of higher solute concentration—in the soil—to the area of lower solute concentration - in the plant.

There are three fundamental ways plants uptake nutrients through the root:

1. Simple diffusion occurs when a nonpolar molecule, such as O_2, CO_2, and NH_3 follows a concentration gradient, moving passively through the cell lipid bilayer membrane without the use of transport proteins.

2. Facilitated diffusion is the rapid movement of solutes or ions following a concentration gradient, facilitated by transport proteins.

3. Active transport is the uptake by cells of ions or molecules against a concentration gradient; this requires an energy source, usually ATP, to power molecular pumps that move the ions or molecules through the membrane.

Nutrients can be moved within plants to where they are most needed. For example, a plant will try to supply more nutrients to its younger leaves than to its older ones. When nutrients are mobile within the plant, symptoms of any deficiency become apparent first on the older leaves. However, not all nutrients are equally mobile. Nitrogen, phosphorus, and potassium are mobile nutrients while the others have varying degrees of mobility. When a less-mobile nutrient is deficient, the younger leaves suffer because the nutrient does not move up to them but stays in the older leaves. This phenomenon is helpful in determining which nutrients a plant may be lacking.

Many plants engage in symbiosis with microorganisms. Two important types of these relationship are

1. with bacteria such as rhizobia, that carry out biological nitrogen fixation, in which atmospheric nitrogen (N_2) is converted into ammonium (NH_4^+); and

2. with mycorrhizal fungi, which through their association with the plant roots help to create a larger effective root surface area. Both of these mutualistic relationships enhance nutrient uptake.

Though nitrogen is plentiful in the Earth's atmosphere, relatively few plants harbour nitrogen-fixing bacteria, so most plants rely on nitrogen compounds present in the soil to support their growth. These can be supplied by mineralization of soil organic matter or added plant residues, nitrogen fixing bacteria, animal waste, through the breaking of triple bonded N_2 molecules by lightening strikes or through the application of fertilizers.

Functions of Nutrients

At least 17 elements are known to be essential nutrients for plants. In relatively large amounts, the soil supplies nitrogen, phosphorus, potassium, calcium, magnesium, and sulfur; these are often called the macronutrients. In relatively small amounts, the soil supplies iron, manganese, boron, molybdenum, copper, zinc, chlorine, and cobalt, the so-called micronutrients. Nutrients must be available not only in sufficient amounts but also in appropriate ratios.

Plant nutrition is a difficult subject to understand completely, partially because of the variation between different plants and even between different species or individuals of a given clone. Elements present at low levels may cause deficiency symptoms, and toxicity is possible at levels that are too high. Furthermore, deficiency of one element may present as symptoms of toxicity from another element, and vice versa. An abundance of one nutrient may cause a deficiency of another nutrient. For example, K^+ uptake can be influenced by the amount of NH_4^+ available.

Although nitrogen is plentiful in the Earth's atmosphere, relatively few plants engage in nitrogen fixation (conversion of atmospheric nitrogen to a biologically useful form). Most plants, therefore, require nitrogen compounds to be present in the soil in which they grow.

Carbon and oxygen are absorbed from the air while other nutrients are absorbed from the soil. Green plants obtain their carbohydrate supply from the carbon dioxide in the air by the process of photosynthesis. Each of these nutrients is used in a different place for a different essential function.

Macronutrients (Derived from air and Water)

Carbon

Carbon forms the backbone of most plant biomolecules, including proteins, starches and cellulose. Carbon is fixed through photosynthesis; this converts carbon dioxide from the air into carbohydrates which are used to store and transport energy within the plant.

Hydrogen

Hydrogen also is necessary for building sugars and building the plant. It is obtained almost entirely from water. Hydrogen ions are imperative for a proton gradient to help drive the electron transport chain in photosynthesis and for respiration.

Oxygen

Oxygen is a component of many organic and inorganic molecules within the plant, and is acquired in many forms. These include: O_2 and CO_2 (mainly from the air via leaves) and H_2O, NO_3^-, $H_2PO_4^-$ and SO_4^{2-} (mainly from the soil water via roots). Plants produce oxygen gas (O_2) along with glucose during photosynthesis but then require O_2 to undergo aerobic cellular respiration and break down this glucose to produce ATP.

Macronutrients (Primary)

Nitrogen

Nitrogen is a major constituent of several of the most important plant substances. For example, nitrogen compounds comprise 40% to 50% of the dry matter of protoplasm, and it is a constituent of amino acids, the building blocks of proteins. It is also an essential constituent of chlorophyll. Nitrogen deficiency most often results in stunted growth, slow growth, and chlorosis. Nitrogen deficient plants will also exhibit a purple appearance on the stems, petioles and underside of leaves from an accumulation of anthocyanin pigments. Most of the nitrogen taken up by plants is from the soil in the forms of NO_3^-, although in acid environments such as boreal forests where nitrification is less likely to occur, ammonium NH_4^+ is more likely to be the dominating source of nitrogen. Amino acids and proteins can only be built from NH_4^+, so NO_3^- must be reduced. In many agricultural settings, nitrogen is the limiting nutrient for rapid growth. Nitrogen is transported via the xylem from the roots to the leaf canopy as nitrate ions, or in an organic form, such as amino acids or amides. Nitrogen can also be transported in the phloem sap as amides, amino acids and ureides; it is therefore mobile within the plant, and the older leaves exhibit chlorosis and necrosis earlier than the younger leaves.

There is an abundant supply of nitrogen in the earth's atmosphere — N_2 gas comprises nearly 79% of air. However, N_2 is unavailable for use by most organisms because there is a triple bond between the two nitrogen atoms in the molecule, making it almost inert. In order for nitrogen to be used for growth it must be "fixed" (combined) in the form of ammonium (NH_4) or nitrate (NO_3) ions. The weathering of rocks releases these ions so slowly that it has a negligible effect on the availability of fixed nitrogen. Therefore, nitrogen is often the limiting factor for growth and biomass production in all environments where there is a suitable climate and availability of water to support life.

Nitrogen enters the plant largely through the roots. A "pool" of soluble nitrogen accumulates. Its composition within a species varies widely depending on several factors, including day length, time of day, night temperatures, nutrient deficiencies, and nutrient imbalance. Short day length promotes asparagine formation, whereas glutamine is produced under long day regimes. Darkness favors protein breakdown accompanied by high asparagine accumulation. Night temperature modifies the effects due to night length, and soluble nitrogen tends to accumulate owing to retarded synthesis and breakdown of proteins. Low night temperature conserves glutamine; high night temperature increases accumulation of asparagine because of breakdown. Deficiency of K accentuates differences between long- and short-day plants. The pool of soluble nitrogen is much smaller than in well-nourished plants when N and P are deficient since uptake of nitrate and further reduction and conversion of N to organic forms is restricted more than is protein synthesis. Deficiencies of Ca, K, and S affect the conversion of organic N to protein more than uptake and reduction. The size of the pool of soluble N is no guide *per se* to growth rate, but the size of the pool in relation to total N might be a useful ratio in this regard. Nitrogen availability in the rooting medium also affects the size and structure of tracheids formed in the long lateral roots of white spruce (Krasowski and Owens 1999).

Microorganisms have a central role in almost all aspects of nitrogen availability, and therefore for life support on earth. Some bacteria can convert N_2 into ammonia by the process termed *nitrogen fixation*; these bacteria are either free-living or form symbiotic associations with plants or other organisms (e.g., termites, protozoa), while other bacteria bring about transformations of ammonia to nitrate, and of nitrate to N_2 or other nitrogen gases. Many bacteria and fungi degrade organic matter, releasing fixed nitrogen for reuse by other organisms. All these processes contribute to the nitrogen cycle.

Phosphorus

Like nitrogen, phosphorus is involved with many vital plant processes. Within a plant, it is present mainly as a structural component of the nucleic acids: deoxyribonucleic acid (DNA) and ribonucleic acid (RNA), as well as a constituent of fatty phospholipids, that are important in membrane development and function. It is present in both organic and inorganic forms, both of which are readily translocated within the plant. All energy transfers in the cell are critically dependent on phosphorus. As with all living things, phosphorus is part of the Adenosine triphosphate (ATP), which is of immediate use in all processes that require energy with the cells. Phosphorus can also be used to modify the activity of various enzymes by phosphorylation, and is used for cell signaling. Phosphorus is concentrated at the most actively growing points of a plant and stored within seeds in anticipation of their germination. Phosphorus is most commonly found in the soil in the form of polyprotic phosphoric acid (H_3PO_4), but is taken up most readily in the form of $H_2PO_4^-$. Phosphorus is available to plants in limited quantities in most soils because it is released very slowly from insoluble phosphates and is rapidly fixed once again. Under most environmental conditions it is the element that limits growth because of this constriction and due to its high demand by plants and microorganisms. Plants can increase phosphorus uptake by a mutualism with mycorrhiza. A Phosphorus deficiency in plants is characterized by an intense green coloration or reddening in leaves due to lack of chlorophyll. If the plant is experiencing high phosphorus deficiencies the leaves may become denatured and show signs of death. Occasionally the leaves may appear purple from an accumulation of anthocyanin. Because phosphorus is a mobile nutrient, older leaves will show the first signs of deficiency.

On some soils, the phosphorus nutrition of some conifers, including the spruces, depends on the ability of mycorrhizae to take up, and make soil phosphorus available to the tree, hitherto unobtainable to the non-mycorrhizal root. Seedling white spruce, greenhouse-grown in sand testing negative for phosphorus, were very small and purple for many months until spontaneous mycorrhizal inoculation, the effect of which was manifested by a greening of foliage and the development of vigorous shoot growth.

Phosphorus deficiency can produce symptoms similar to those of nitrogen deficiency, but as noted by Russel: "Phosphate deficiency differs from nitrogen deficiency in being extremely difficult to diagnose, and crops can be suffering from extreme starvation without there being any obvious signs that lack of phosphate is the cause". Russell's observation applies to at least some coniferous seedlings, but Benzian found that although response to phosphorus in very acid forest tree nurseries in England was consistently high, no species (including Sitka spruce) showed any visible symptom of deficiency other than a slight lack of lustre. Phosphorus levels have to be exceedingly low before visible symptoms appear in such seedlings. In sand culture at 0 ppm phosphorus, white spruce seedlings were very small and tinted deep purple; at 0.62 ppm, only the smallest seedlings were deep purple; at 6.2 ppm, the seedlings were of good size and color.

It is useful to apply a high phosphorus content fertilizer, such as bone meal, to perennials to help with successful root formation.

Potassium

Unlike other major elements, potassium does not enter into the composition of any of the important plant constituents involved in metabolism, but it does occur in all parts of plants in substantial amounts. It seems to be of particular importance in leaves and at growing points. Potassium is outstanding among the nutrient elements for its mobility and solubility within plant tissues. Processes involving potassium include the formation of carbohydrates and proteins, the regulation of internal plant moisture, as a catalyst and condensing agent of complex substances, as an accelerator of enzyme action, and as contributor to photosynthesis, especially under low light intensity.

Potassium regulates the opening and closing of the stomata by a potassium ion pump. Since stomata are important in water regulation, potassium regulates water loss from the leaves and increases drought tolerance. Potassium deficiency may cause necrosis or interveinal chlorosis. The potassium ion (K^+) is highly mobile and can aid in balancing the anion (negative) charges within the plant. Potassium helps in fruit coloration, shape and also increases its brix. Hence, quality fruits are produced in potassium-rich soils. Potassium serves as an activator of enzymes used in photosynthesis and respiration. Potassium is used to build cellulose and aids in photosynthesis by the formation of a chlorophyll precursor. Potassium deficiency may result in higher risk of pathogens, wilting, chlorosis, brown spotting, and higher chances of damage from frost and heat.

When soil-potassium levels are high, plants take up more potassium than needed for healthy growth. The term *luxury consumption* has been applied to this. When potassium is moderately deficient, the effects first appear in the older tissues, and from there progress towards the growing points. Acute deficiency severely affects growing points, and die-back commonly occurs. Symptoms of potassium deficiency in white spruce include: browning and death of needles (chlorosis); reduced growth in height and diameter; impaired retention of needles; and reduced needle length. A relationship between potassium nutrition and cold resistance has been found in several tree species, including two species of spruce.

Macronutrients (Secondary and Tertiary)

Sulfur

Sulfur is a structural component of some amino acids and vitamins, and is essential in the manufacturing of chloroplasts. Sulphur is also found in the iron-sulphur complexes of the electron transport chains in photosynthesis. Sulphate ions are mobile and its deficiency, therefore, affects older tissues first. Symptoms of deficiency include yellowing of leaves and stunted growth.

Calcium

Calcium regulates transport of other nutrients into the plant and is also involved in the activation of certain plant enzymes. Calcium deficiency results in stunting. This nutrient is involved in photosynthesis and plant structure. Blossom end rot is also a result of inadequate calcium.

Another common symptom of calcium deficiency in leaves is the curling of the leaf towards the veins or center of the leaf. Many times this can also have a blackened appearance Calcium has been found to have a positive effect in combating salinity in soils. It has been shown to ameliorate the negative effects that salinity has such as reduced water usage of plants. Calcium in plants occurs chiefly in the leaves, with lower concentrations in seeds, fruits, and roots. A major function is as a constituent of cell walls. When coupled with certain acidic compounds of the jelly-like pectins of the middle lamella, calcium forms an insoluble salt. It is also intimately involved in meristems, and is particularly important in root development, with roles in cell division, cell elongation, and the detoxification of hydrogen ions. Other functions attributed to calcium are; the neutralization of organic acids; inhibition of some potassium-activated ions; and a role in nitrogen absorption. A notable feature of calcium-deficient plants is a defective root system. Roots are usually affected before above-ground parts.

Magnesium

The outstanding role of magnesium in plant nutrition is as a constituent of the chlorophyll molecule. As a carrier, it is also involved in numerous enzyme reactions as an effective activator, in which it is closely associated with energy-supplying phosphorus compounds. Magnesium is very mobile in plants, and, like potassium, when deficient is translocated from older to younger tissues, so that signs of deficiency appear first on the oldest tissues and then spread progressively to younger tissues.

Micro-nutrients

Plants are able sufficiently to accumulate most trace elements. Some plants are sensitive indicators of the chemical environment in which they grow (Dunn 1991), and some plants have barrier mechanisms that exclude or limit the uptake of a particular element or ion species, e.g., alder twigs commonly accumulate molybdenum but not arsenic, whereas the reverse is true of spruce bark (Dunn 1991). Otherwise, a plant can integrate the geochemical signature of the soil mass permeated by its root system together with the contained groundwaters. Sampling is facilitated by the tendency of many elements to accumulate in tissues at the plant's extremities.

Iron

Iron is necessary for photosynthesis and is present as an enzyme cofactor in plants. Iron deficiency can result in interveinal chlorosis and necrosis. Iron is not a structural part of chlorophyll but very much essential for its synthesis. Copper deficiency can be responsible for promoting an iron deficiency. It helps in the electron transport of plant.

Molybdenum

Molybdenum is a cofactor to enzymes important in building amino acids and is involved in nitrogen metabolism. Molybdenum is part of the nitrate reductase enzyme (needed for the reduction of nitrate) and the nitrogenase enzyme (required for biological nitrogen fixation).

Boron

Boron is found in the highly insoluble mineral, tourmaline (a crystalline boron silicate mineral). It is absorbed by plants in the form of the anion BO_3^-. It is available to plants in moderately soluble mineral forms of Ca, Mg and Na borates and the highly soluble form of organic compounds. Concentration in soil must, in general, be below 5 ppm in the soil water solution, above that toxicity results. Its availability in soils ranges from 20 to 200 pounds per acre in the first eight inches, of which less than 5% is available. It is available to plants over a range of pH, from 5.0 to 7.5. It is mobile in the soil, hence, it is prone to leaching. Leaching removes substantial amounts of boron in sandy soil, but little in fine silt or clay soil. Boron's fixation to those minerals at high pH can render boron unavailable, while low pH frees the fixed boron, leaving it prone to leaching in wet climates. It precipitates with other minerals in the form of borax in which form it was first used over 400 years ago as a soil supplement. Decomposition of organic material causes boron to be deposited in the topmost soil layer; organic forms of boron are more soluble than their mineral form, hence are more available in the top few inches. When soil dries it can cause a precipitous drop in the availability of boron to plants as the plants cannot draw nutrients from that desiccated layer. Hence, boron deficiency diseases appear in dry weather.

Boron has many functions within a plant: it affects flowering and fruiting, pollen germination, cell division, and active salt absorption. The metabolism of amino acids and proteins, carbohydrates, calcium, and water are strongly affected by boron. Many of those listed functions may be embodied by its function in moving the highly polar sugars through cell membranes by reducing their polarity and hence the energy needed to pass the sugar. If sugar cannot pass to the fastest growing parts rapidly enough, those parts die. Boron is relatively immobile within a plant suggesting that the molecule is fixed to the points in the membrane where they facilitate sugar transport.

Boron is not relocatable in the plant via the phloem. It must be supplied to the growing parts via the xylem. Foliar sprays affect only those parts sprayed, which may be insufficient for the fastest growing parts, and is very temporary.

Boron is essential for the proper forming and strengthening of cell walls. Lack of boron results in short thick cells producing stunted fruiting bodies and roots. Calcium to boron ratio must be maintained in a narrow range for normal plant growth. For alfalfa, that calcium to boron ratio must be from 80:1 to 600:1. Boron deficiency appears at 800:1 and higher. For alfalfa, similar ratios exist for magnesium, copper, nitrogen and potassium. Boron levels within plants differ with plant species and range from 2.3 p.p.m for barley to 94.7 p.p.m for poppy . Lack of boron causes failure of calcium metabolism which produces hollow heart in beets and peanuts.

Inadequate amounts of boron affect many agricultural crops, legume forage crops most strongly. Of the micronutrients, boron deficiencies are second most common after zinc. Deficiencies of boron when soil is cropped are common and require the application of mineral supplement; one ton of alfalfa hay carries with it one ounce of boron, 100 bushels of peaches 4 ounces. Deficiency results in the death of the terminal growing points. Symptoms first appear as stunted growth, then to cellular changes, which leads to physical changes, and finally death of the plant.

Boron supplements derive from dry lake bed deposits such as those in Death Valley, USA, in the form of sodium tetraborate (borax), from which less soluble calcium borate is made. Foliar sprays are used on fruit crop trees in soils of high alkalinity. Boron is often applied to fields as a contaminant in other soil amendments but is not generally adequate to make up the rate of loss by cropping. The rates of application of borate to produce an adequate alfalfa crop range from 15 pounds per acre for a sandy-silt, acidic soil of low organic matter, to 60 pounds per acre for a soil with high organic matter, high cation exchange capacity and high pH.

Boron concentration in soil water solution higher than one ppm is toxic to most plants. Toxic concentrations within plants are 10 to 50 ppm for small grains and 200 ppm in boron-tolerant crops such as sugar beets, rutabaga, cucumbers, and conifers. Toxic soil conditions are generally limited to arid regions or can be caused by underground borax deposits in contact with water or volcanic gases dissolved in percolating water. Application rates should be limited to a few pounds per acre in a test plot to determine if boron is needed generally. Otherwise, testing for boron levels in plant material is required to determine remedies. Excess boron can be removed by irrigation and assisted by application of elemental sulfur to lower the pH and increase boron's solubility. Application of calcium will increase soil alkalinity, causing boron to fix on the mineral soil component and remove some fraction, thereby reducing boron toxicity.

Boron deficiencies must be detected by analysis of plant material to apply a correction before the obvious symptoms appear, after which it is too late to prevent crop loss. Strawberries deficient in boron will produce lumpy fruit; apricots will not blossom or, if they do, will not fruit or will drop their fruit depending on the level of boron deficit. Broadcast of boron supplements is effective and long term; a foliar spray is immediate but must be repeated.

Boron is an essential element for the health of animals which derive their boron from plant material.

Copper

Copper is important for photosynthesis. Symptoms for copper deficiency include chlorosis.It is involved in many enzyme processes; necessary for proper photosynthesis; involved in the manufacture of lignin (cell walls) and involved in grain production. It is also hard to find in some soil conditions.

Manganese

Manganese is necessary for photosynthesis, including the building of chloroplasts. Manganese deficiency may result in coloration abnormalities, such as discolored spots on the foliage.

Sodium

Sodium is involved in the regeneration of phosphoenolpyruvate in CAM and C_4 plants. Sodium can potentially replace potassium's regulation of stomatal opening and closing.

Essentiality of sodium:

- Essential for C_4 plants rather C_3

- Substitution of K by Na: Plants can be classified into four groups:

    1. Group A—a high proportion of K can be replaced by Na and stimulate the growth, which cannot be achieved by the application of K

    2. Group B—specific growth responses to Na are observed but they are much less distinct

    3. Group C—Only minor substitution is possible and Na has no effect

    4. Group D—No substitution occurs

- Stimulate the growth—increase leaf area and stomata. Improves the water balance

- Na functions in metabolism

    1. C_4 metabolism

    2. Impair the conversion of pyruvate to phosphoenol-pyruvate

    3. Reduce the photosystem II activity and ultrastructural changes in mesophyll chloroplast

- Replacing K functions

    1. Internal osmoticum

    2. Stomatal function

    3. Photosynthesis

    4. Counteraction in long distance transport

    5. Enzyme activation

- Improves the crop quality e.g. improves the taste of carrots by increasing sucrose

Zinc

Zinc is required in a large number of enzymes and plays an essential role in DNA transcription. A typical symptom of zinc deficiency is the stunted growth of leaves, commonly known as "little leaf" and is caused by the oxidative degradation of the growth hormone auxin.

Nickel

In higher plants, nickel is absorbed by plants in the form of Ni^{2+} ion. Nickel is essential for activation of urease, an enzyme involved with nitrogen metabolism that is required to process urea. Without nickel, toxic levels of urea accumulate, leading to the formation of necrotic lesions. In lower plants, nickel activates several enzymes involved in a variety of processes, and can substitute for zinc and iron as a cofactor in some enzymes.

Chlorine

Chlorine, as compounded chloride, is necessary for osmosis and ionic balance; it also plays a role in photosynthesis.

Cobalt

Cobalt has proven to be beneficial to at least some plants although it does not appear to be essential for most species. It has, however, been shown to be essential for nitrogen fixation by the nitrogen-fixing bacteria associated with legumes and other plants.

Aluminum

Aluminum is one of the few elements capable of making soil more acidic. This is achieved by aluminum taking hydroxide ions out of water, leaving hydrogen ions behind. As a result, the soil is more acidic, which makes it unlivable for many plants. Another consequence of aluminum in soils is aluminum toxicity, which inhibits root growth.

- Tea has a high tolerance for aluminum (Al) toxicity and the growth is stimulated by Al application. The possible reason is the prevention of Cu, Mn or P toxicity effects.

- There have been reports that Al may serve as a fungicide against certain types of root rot.

Silicon

Silicon is not considered an essential element for plant growth and development. It is always found in abundance in the environment and hence if needed it is available. It is found in the structures of plants and improves the health of plants.

In plants, silicon has been shown in experiments to strengthen cell walls, improve plant strength, health, and productivity. There have been studies showing evidence of silicon improving drought and frost resistance, decreasing lodging potential and boosting the plant's natural pest and disease fighting systems. Silicon has also been shown to improve plant vigor and physiology by improving root mass and density, and increasing above ground plant biomass and crop yields. Silicon is currently under consideration by the Association of American Plant Food Control Officials (AAPFCO) for elevation to the status of a "plant beneficial substance".

Vanadium

Vanadium may be required by some plants, but at very low concentrations. It may also be substituting for molybdenum.

Selenium

Selenium is probably not essential for flowering plants, but it can be beneficial; it can stimulate plant growth, improve tolerance of oxidative stress, and increase resistance to pathogens and herbivory.

Selenium is, however, an essential mineral element for animal (including human) nutrition and selenium deficiencies are known to occur when food or animal feed is grown on selenium-deficient soils. The use of inorganic selenium fertilizers can increase selenium concentrations in edible crops and animal diets thereby improving animal health.

Nutrient Deficiency

The effect of a nutrient deficiency can vary from a subtle depression of growth rate to obvious stunting, deformity, discoloration, distress, and even death. Visual symptoms distinctive enough to be useful in identifying a deficiency are rare. Most deficiencies are multiple and moderate. However, while a deficiency is seldom that of a single nutrient, nitrogen is commonly the nutrient in shortest supply.

Chlorosis of foliage is not always due to mineral nutrient deficiency. Solarization can produce superficially similar effects, though mineral deficiency tends to cause premature defoliation, whereas solarization does not, nor does solarization depress nitrogen concentration.

Nutrient Status of Plants

Nutrient status (mineral nutrient and trace element composition, also called ionome and nutrient profile) of plants are commonly portrayed by tissue elementary analysis. Interpretation of the results of such studies, however, has been controversial. During the last decades the nearly two-century-old "law of minimum" or "Liebig's law" (that states that plant growth is controlled not by the total amount of resources available, but by the scarcest resource) has been replaced by several mathematical approaches that use different models in order to take the interactions between the individual nutrients into account. The latest developments in this field are based on the fact that the nutrient elements (and compounds) do not act independently from each other; Baxter, 2015, because there may be direct chemical interactions between them or they may influence each other's uptake, translocation, and biological action via a number of mechanisms as exemplified for the case of ammonia.

Plant Nutrition in Agricultural Systems

Hydroponics

Hydroponics is a method for growing plants in a water-nutrient solution without the use of nutrient-rich soil. It allows researchers and home gardeners to grow their plants in a controlled environment. The most common solution is the Hoagland solution, developed by D. R. Hoagland in 1933. The solution consists of all the essential nutrients in the correct proportions necessary for most plant growth. An aerator is used to prevent an anoxic event or hypoxia. Hypoxia can affect nutrient uptake of a plant because, without oxygen present, respiration becomes inhibited within the root cells. The nutrient film technique is a hydroponic technique in which the roots are not fully submerged. This allows for adequate aeration of the roots, while a "film" thin layer of nutrient-rich water is pumped through the system to provide nutrients and water to the plant.

Plant Hormone

Plant hormones (also known as phytohormones) are chemicals that regulate plant growth. In the United Kingdom, these are termed 'plant growth substances'.

Plant hormones are signal molecules produced within the plant, and occur in extremely low concentrations. Hormones regulate cellular processes in targeted cells locally and, moved to other locations, in other functional parts of the plant. Hormones also determine the formation of flowers, stems, leaves, the shedding of leaves, and the development and ripening of fruit. Plants, unlike animals, lack glands that produce and secrete hormones. Instead, each cell is capable of producing hormones. Plant hormones shape the plant, affecting seed growth, time of flowering, the sex of flowers, senescence of leaves, and fruits. They affect which tissues grow upward and which grow downward, leaf formation and stem growth, fruit development and ripening, plant longevity, and even plant death. Hormones are vital to plant growth, and, lacking them, plants would be mostly a mass of undifferentiated cells. So they are also known as growth factors or growth hormones. The term 'Phytohormone' was coined by Thimann in 1948.

Phytohormones are found not only in higher plants but in algae, showing similar functions, and in microorganisms, such as unicellular fungi and bacteria, but in these cases they play no hormonal or other immediate physiological role in the producing organism and can, thus, be regarded as secondary metabolites.

Characteristics

The word hormone is derived from Greek, meaning *set in motion*. Plant hormones affect gene expression and transcription levels, cellular division, and growth. They are naturally produced within plants, though very similar chemicals are produced by fungi and bacteria that can also affect plant growth. A large number of related chemical compounds are synthesized by humans. They are used to regulate the growth of cultivated plants, weeds, and in vitro-grown plants and plant cells; these manmade compounds are called Plant Growth Regulators or PGRs for short. Early in the study of plant hormones, "phytohormone" was the commonly used term, but its use is less widely applied now.

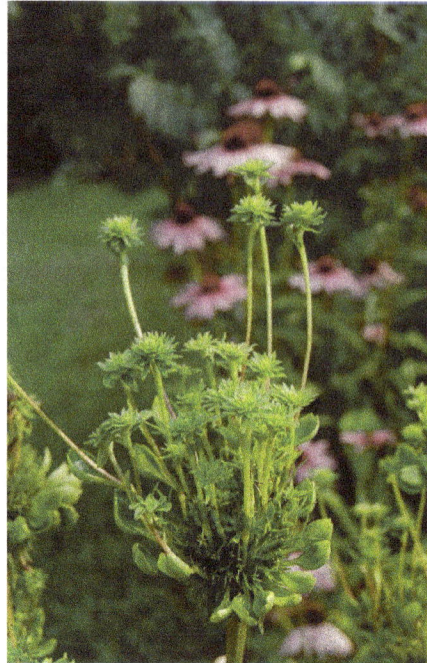

Phyllody on a purple coneflower (*Echinacea purpurea*), a plant development abnormality where leaf-like
structures replace flower organs. It can be caused by hormonal imbalance, among other reasons.

Plant hormones are not nutrients, but chemicals that in small amounts promote and influence the
growth, development, and differentiation of cells and tissues. The biosynthesis of plant hormones
within plant tissues is often diffuse and not always localized. Plants lack glands to produce and
store hormones, because, unlike animals — which have two circulatory systems (lymphatic and
cardiovascular) powered by a heart that moves fluids around the body — plants use more passive
means to move chemicals around their bodies. Plants utilize simple chemicals as hormones, which
move more easily through their tissues. They are often produced and used on a local basis within
the plant body. Plant cells produce hormones that affect even different regions of the cell producing the hormone.

Hormones are transported within the plant by utilizing four types of movements. For localized
movement, cytoplasmic streaming within cells and slow diffusion of ions and molecules between
cells are utilized. Vascular tissues are used to move hormones from one part of the plant to another; these include sieve tubes or phloem that move sugars from the leaves to the roots and flowers,
and xylem that moves water and mineral solutes from the roots to the foliage.

Not all plant cells respond to hormones, but those cells that do are programmed to respond at
specific points in their growth cycle. The greatest effects occur at specific stages during the cell's
life, with diminished effects occurring before or after this period. Plants need hormones at very
specific times during plant growth and at specific locations. They also need to disengage the effects
that hormones have when they are no longer needed. The production of hormones occurs very
often at sites of active growth within the meristems, before cells have fully differentiated. After
production, they are sometimes moved to other parts of the plant, where they cause an immediate
effect; or they can be stored in cells to be released later. Plants use different pathways to regulate
internal hormone quantities and moderate their effects; they can regulate the amount of chemicals
used to biosynthesize hormones. They can store them in cells, inactivate them, or cannibalise al-

ready-formed hormones by conjugating them with carbohydrates, amino acids, or peptides. Plants can also break down hormones chemically, effectively destroying them. Plant hormones frequently regulate the concentrations of other plant hormones. Plants also move hormones around the plant diluting their concentrations.

The concentration of hormones required for plant responses are very low (10^{-6} to 10^{-5} mol/L). Because of these low concentrations, it has been very difficult to study plant hormones, and only since the late 1970s have scientists been able to start piecing together their effects and relationships to plant physiology. Much of the early work on plant hormones involved studying plants that were genetically deficient in one or involved the use of tissue-cultured plants grown *in vitro* that were subjected to differing ratios of hormones, and the resultant growth compared. The earliest scientific observation and study dates to the 1880s; the determination and observation of plant hormones and their identification was spread-out over the next 70 years.

Classes of Plant Hormones

In general, it is accepted that there are five major classes of plant hormones, some of which are made up of many different chemicals that can vary in structure from one plant to the next. The chemicals are each grouped together into one of these classes based on their structural similarities and on their effects on plant physiology. Other plant hormones and growth regulators are not easily grouped into these classes; they exist naturally or are synthesized by humans or other organisms, including chemicals that inhibit plant growth or interrupt the physiological processes within plants. Each class has positive as well as inhibitory functions, and most often work in tandem with each other, with varying ratios of one or more interplaying to affect growth regulation.

The five major classes are:

Abscisic Acid Hormone

Abscisic acid (also called ABA) is one of the most important plant growth regulators. It was discovered and researched under two different names before its chemical properties were fully known, it was called *dormin* and *abscicin II*. Once it was determined that the two compounds are the same, it was named abscisic acid. The name "abscisic acid" was given because it was found in high concentrations in newly abscissed or freshly fallen leaves.

This class of PGR is composed of one chemical compound normally produced in the leaves of plants, originating from chloroplasts, especially when plants are under stress. In general, it acts as an inhibitory chemical compound that affects bud growth, and seed and bud dormancy. It mediates changes within the apical meristem, causing bud dormancy and the alteration of the last set of leaves into protective bud covers. Since it was found in freshly abscissed leaves, it was thought to play a role in the processes of natural leaf drop, but further research has disproven this. In plant species from temperate parts of the world, it plays a role in leaf and seed dormancy by inhibiting growth, but, as it is dissipated from seeds or buds, growth begins. In other plants, as ABA levels decrease, growth then commences as gibberellin levels increase. Without ABA, buds and seeds would start to grow during warm periods in winter and be killed when it froze again. Since ABA dissipates slowly from the tissues and its effects take time to be offset by other plant hormones, there is a delay in physiological pathways that provide some protection from premature growth. It accumulates within seeds during fruit maturation, preventing seed germination within the fruit, or seed germination before winter. Abscisic acid's effects are degraded within plant tissues during cold temperatures or by its removal by water washing in out of the tissues, releasing the seeds and buds from dormancy.

In plants under water stress, ABA plays a role in closing the stomata. Soon after plants are water-stressed and the roots are deficient in water, a signal moves up to the leaves, causing the formation of ABA precursors there, which then move to the roots. The roots then release ABA, which is translocated to the foliage through the vascular system and modulates the potassium and sodium uptake within the guard cells, which then lose turgidity, closing the stomata. ABA exists in all parts of the plant and its concentration within any tissue seems to mediate its effects and function as a hormone; its degradation, or more properly catabolism, within the plant affects metabolic reactions and cellular growth and production of other hormones. Plants start life as a seed with high ABA levels. Just before the seed germinates, ABA levels decrease; during germination and early growth of the seedling, ABA levels decrease even more. As plants begin to produce shoots with fully functional leaves, ABA levels begin to increase, slowing down cellular growth in more "mature" areas of the plant. Stress from water or predation affects ABA production and catabolism rates, mediating another cascade of effects that trigger specific responses from targeted cells. Scientists are still piecing together the complex interactions and effects of this and other phytohormones.

Auxins

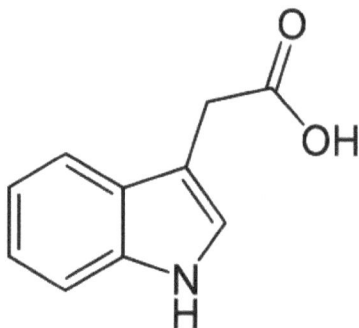

The auxin indole-3-acetic acid

Auxins are compounds that positively influence cell enlargement, bud formation and root initiation. They also promote the production of other hormones and in conjunction with cytokinins, they control the growth of stems, roots, and fruits, and convert stems into flowers. Auxins were the first class of growth regulators discovered. They affect cell elongation by altering cell wall plasticity. They stimulate cambium, a subtype of meristem cells, to divide and in stems cause secondary xylem to differentiate. Auxins act to inhibit the growth of buds lower down the stems (apical dominance), and also to promote lateral and adventitious root development and growth. Leaf abscission is initiated by the growing point of a plant ceasing to produce auxins. Auxins in seeds regulate specific protein synthesis, as they develop within the flower after pollination, causing the flower to develop a fruit to contain the developing seeds. Auxins are toxic to plants in large concentrations; they are most toxic to dicots and less so to monocots. Because of this property, synthetic auxin herbicides including 2,4-D and 2,4,5-T have been developed and used for weed control. Auxins, especially 1-Naphthaleneacetic acid (NAA) and Indole-3-butyric acid (IBA), are also commonly applied to stimulate root growth when taking cuttings of plants. The most common auxin found in plants is indole-3-acetic acid or IAA. The correlation of auxins and cytokinins in the plants is a constant (A/C = const.).

Cytokinins

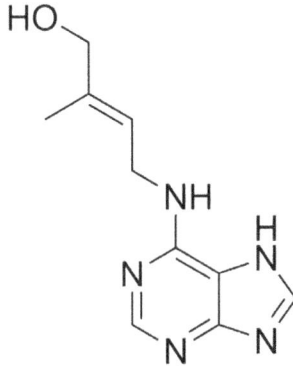

The cytokinin zeatin, the name is derived from *Zea* maize, in which it was first discovered in immature kernels.

Cytokinins or CKs are a group of chemicals that influence cell division and shoot formation. They were called kinins in the past when the first cytokinins were isolated from yeast cells. They also help delay senescence of tissues, are responsible for mediating auxin transport throughout the plant, and affect internodal length and leaf growth. They have a highly synergistic effect in concert with auxins, and the ratios of these two groups of plant hormones affect most major growth periods during a plant's lifetime. Cytokinins counter the apical dominance induced by auxins; they in conjunction with ethylene promote abscission of leaves, flower parts, and fruits. The correlation of auxins and cytokinins in the plants is a constant (A/C = const.).

Ethylene

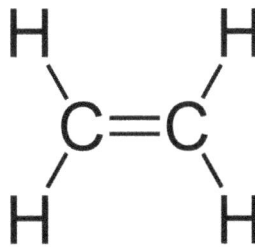

Ethylene

Ethylene is a gas that forms through the breakdown of methionine, which is in all cells. Ethylene has very limited solubility in water and does not accumulate within the cell but diffuses out of the cell and escapes out of the plant. Its effectiveness as a plant hormone is dependent on its rate of production versus its rate of escaping into the atmosphere. Ethylene is produced at a faster rate in rapidly growing and dividing cells, especially in darkness. New growth and newly germinated seedlings produce more ethylene than can escape the plant, which leads to elevated amounts of ethylene, inhibiting leaf expansion. As the new shoot is exposed to light, reactions by phytochrome in the plant's cells produce a signal for ethylene production to decrease, allowing leaf expansion. Ethylene affects cell growth and cell shape; when a growing shoot hits an obstacle while underground, ethylene production greatly increases, preventing cell elongation and causing the stem to swell. The resulting thicker stem can exert more pressure against the object impeding its path to the surface. If the shoot does not reach the surface and the ethylene stimulus becomes prolonged, it affects the stem's natural geotropic response, which is to grow upright, allowing it to grow around an object. Studies seem to indicate that ethylene affects stem diameter and height: When stems of trees are subjected to wind, causing lateral stress, greater ethylene production occurs, resulting in thicker, more sturdy tree trunks and branches. Ethylene affects fruit-ripening: Normally, when the seeds are mature, ethylene production increases and builds-up within the fruit, resulting in a climacteric event just before seed dispersal. The nuclear protein Ethylene Insensitive2 (EIN2) is regulated by ethylene production, and, in turn, regulates other hormones including ABA and stress hormones.

Gibberellins

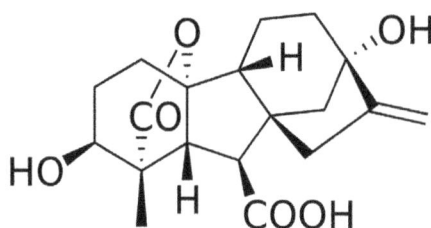

Gibberellin A1

Main function: initiate mobilization of storage materials in seeds during germination, cause elongation of stems, stimulate bolting in biennials stimulate pollen tube growth.

Gibberellins, or GAs, include a large range of chemicals that are produced naturally within plants and by fungi. They were first discovered when Japanese researchers, including Eiichi Kurosawa, noticed a chemical produced by a fungus called *Gibberella fujikuroi* that produced abnormal growth in rice plants. Gibberellins are important in seed germination, affecting enzyme production that mobilizes food production used for growth of new cells. This is done by modulating chromosomal transcription. In grain (rice, wheat, corn, etc.) seeds, a layer of cells called the aleurone layer wraps around the endosperm tissue. Absorption of water by the seed causes production of GA. The GA is transported to the aleurone layer, which responds by producing enzymes that break down stored food reserves within the endosperm, which are utilized by the growing seedling. GAs produce bolting of rosette-forming plants, increasing internodal length. They promote flowering, cellular division, and in seeds growth after germination. Gibberellins also reverse the inhibition of shoot growth and dormancy induced by ABA.

Other Known Hormones

Other identified plant growth regulators include:

- Brassinosteroids - are a class of polyhydroxysteroids, a group of plant growth regulators. Brassinosteroids have been recognized as a sixth class of plant hormones, which stimulate cell elongation and division, gravitropism, resistance to stress, and xylem differentiation. They inhibit root growth and leaf abscission. Brassinolide was the first identified brassinosteroid and was isolated from extracts of rapeseed (*Brassica napus*) pollen in 1979.

- Salicylic acid — activates genes in some plants that produce chemicals that aid in the defense against pathogenic invaders.

- Jasmonates — are produced from fatty acids and seem to promote the production of defense proteins that are used to fend off invading organisms. They are believed to also have a role in seed germination, and affect the storage of protein in seeds, and seem to affect root growth.

- Plant peptide hormones — encompasses all small secreted peptides that are involved in cell-to-cell signaling. These small peptide hormones play crucial roles in plant growth and development, including defense mechanisms, the control of cell division and expansion, and pollen self-incompatibility.

- Polyamines — are strongly basic molecules with low molecular weight that have been found in all organisms studied thus far. They are essential for plant growth and development and affect the process of mitosis and meiosis.

- Nitric oxide (NO) — serves as signal in hormonal and defense responses (e.g. stomatal closure, root development, germination, nitrogen fixation, cell death, stress response). NO can be produced by a yet undefined NO synthase, a special type of nitrite reductase, nitrate reductase, mitochondrial cytochrome c oxidase or non enzymatic processes and regulate plant cell organelle functions (e.g. ATP synthesis in chloroplasts and mitochondria).

- Strigolactones - implicated in the inhibition of shoot branching.

- Karrikins - not plant hormones because they are not made by plants, but are a group of plant growth regulators found in the smoke of burning plant material that have the ability to stimulate the germination of seeds

- Triacontanol - a fatty alcohol that acts as a growth stimulant, especially initiating new basal breaks in the rose family. It is found in alfalfa (lucerne), bee's wax, and some waxy leaf cuticles.

Potential Medical Applications

Plant stress hormones activate cellular responses, including cell death, to diverse stress situations in plants. Researchers have found that some plant stress hormones share the ability to adversely affect human cancer cells. For example, sodium salicylate has been found to suppress proliferation of lymphoblastic leukemia, prostate, breast, and melanoma human cancer cells. Jasmonic acid, a plant stress hormone that belongs to the jasmonate family, induced death in lymphoblastic leukemia cells. Methyl jasmonate has been found to induce cell death in a number of cancer cell lines.

Hormones and Plant Propagation

Synthetic plant hormones or PGRs are commonly used in a number of different techniques involving plant propagation from cuttings, grafting, micropropagation, and tissue culture.

The propagation of plants by cuttings of fully developed leaves, stems, or roots is performed by gardeners utilizing auxin as a rooting compound applied to the cut surface; the auxins are taken into the plant and promote root initiation. In grafting, auxin promotes callus tissue formation, which joins the surfaces of the graft together. In micropropagation, different PGRs are used to promote multiplication and then rooting of new plantlets. In the tissue-culturing of plant cells, PGRs are used to produce callus growth, multiplication, and rooting.

Seed Dormancy

Plant hormones affect seed germination and dormancy by acting on different parts of the seed.

Embryo dormancy is characterized by a high ABA:GA ratio, whereas the seed has a high ABA sensitivity and low GA sensitivity. In order to release the seed from this type of dormancy and initiate seed germination, an alteration in hormone biosynthesis and degradation toward a low ABA/GA ratio, along with a decrease in ABA sensitivity and an increase in GA sensitivity, must occur.

ABA controls embryo dormancy, and GA embryo germination. Seed coat dormancy involves the mechanical restriction of the seed coat. This, along with a low embryo growth potential, effectively produces seed dormancy. GA releases this dormancy by increasing the embryo growth potential, and/or weakening the seed coat so the radical of the seedling can break through the seed coat. Different types of seed coats can be made up of living or dead cells, and both types can be influenced by hormones; those composed of living cells are acted upon after seed formation, whereas the seed coats composed of dead cells can be influenced by hormones during the formation of the seed coat. ABA affects testa or seed coat growth characteristics, including thickness, and effects the GA-mediated embryo growth potential. These conditions and effects occur during the formation of the seed, often in response to environmental conditions. Hormones also mediate endosperm dormancy: Endosperm in most seeds is composed of living tissue that can actively respond to hormones generated by the embryo. The endosperm often acts as a barrier to seed germination, playing a part in seed coat dormancy or in the germination process. Living cells respond to and also affect the ABA:GA ratio, and mediate cellular sensitivity; GA thus increases the embryo growth potential and can promote endosperm weakening. GA also affects both ABA-independent and ABA-inhibiting processes within the endosperm.

Photomorphogenesis

In developmental biology, photomorphogenesis is light-mediated development, where plant growth patterns respond to the light spectrum. This is a completely separate process from photosynthesis where light is used as a source of energy. Phytochromes, cryptochromes, and phototropins are photochromic sensory receptors that restrict the photomorphogenic effect of light to the UV-A, UV-B, blue, and red portions of the electromagnetic spectrum.

The photomorphogenesis of plants is often studied by using tightly frequency-controlled light sources to grow the plants. There are at least three stages of plant development where photomorphogenesis occurs: seed germination, seedling development, and the switch from the vegetative to the flowering stage (photoperiodism).

History

Theophrastus of Eresus (371 to 287 BC) may have been the first to write about photomorphogenesis. He described the different wood qualities of fir trees grown in different levels of light, likely the result of the photomorphogenic "shade avoidance effect." In 1686, John Ray wrote "Historia Plantarum" which mentioned the effects of etiolation. Charles Bonnet introduced the term "etiolement" to the scientific literature in 1754 when describing his experiments, commenting that the term was already in use by gardeners.

Developmental Stages Affected

Seed Germination

Light has profound effects on the development of plants. The most striking effects of light are observed when a germinating seedling emerges from the soil and is exposed to light for the first time.

Normally the seedling radicle (root) emerges first from the seed, and the shoot appears as the root becomes established. Later, with growth of the shoot (particularly when it emerges into the light) there is increased secondary root formation and branching. In this coordinated progression of developmental responses are early manifestations of correlative growth phenomena where the root affects the growth of the shoot and vice versa. To a large degree, the growth responses are hormone mediated.

Seedling Development

In the absence of light, plants develop an etiolated growth pattern. Etiolation of the seedling causes it to become elongated, which may facilitate it emerging from the soil.

A seedling that emerges in darkness follows a developmental program known as skotomorphogenesis (dark development), which is characterized by etiolation. Upon exposure to light, the seedling switches rapidly to photomorphogenesis (light development).

There are differences when comparing dark-grown (etiolated) and light-grown (de-etiolated) seedlings

A dicot seedling emerging from the ground displays an apical hook (in the hypocotyl in this case),
a response to dark conditions

Etiolated characteristics:

- Distinct apical hook (dicot) or coleoptile (monocot)
- No leaf growth
- No chlorophyll
- Rapid stem elongation
- Limited radial expansion of stem
- Limited root elongation
- Limited production of lateral roots

De-etiolated Characteristics:

- Apical hook opens or coleoptile splits open
- Leaf growth promoted
- Chlorophyll produced
- Stem elongation suppressed
- Radial expansion of stem
- Root elongation promoted
- Lateral root development accelerated

The developmental changes characteristic of photomorphogenesis shown by de-etiolated seedlings, are induced by light.

Photoperiodism

Some plants rely on light signals to determine when to switch from the vegetative to the flowering stage of plant development. This type of photomorphogenesis is known as photoperiodism and involves using red photoreceptors (phytochromes) to determine the daylength. As a result, photoperiodic plants only start making flowers when the days have reached a "critical daylength," allowing these plants to initiate their flowering period according to the time of year. For example, "long day" plants need long days to start flowering, and "short day" plants need to experience short days before they will start making flowers.

Photoperiodism also has an effect on vegetative growth, including on bud dormancy in perennial plants, though this is not as well-documented as the effect of photoperiodism on the switch to the flowering stage.

Light Receptors for Photomorphogenesis

Typically, plants are responsive to wavelengths of light in the blue, red and far-red regions of the spectrum through the action of several different photosensory systems. The photoreceptors for red and far-red wavelengths are known as phytochromes. There are at least 5 members of the phytochrome family of photoreceptors. There are several blue light photoreceptors known as cryptochromes. The combination of phytochromes and cryptochromes mediate growth and the flowering of plants in response to red light, far-red light, and blue light.

Red/Far-red Light

Plants use phytochrome to detect and respond to red and far-red wavelengths. Phytochromes are signaling proteins that promote photomorphogenesis in response to red light and far-red light. Phytochrome is the only known photoreceptor that absorbs light in the red/far red spectrum of light (600-750 nm) specifically and only for photosensory purposes. Phytochromes are proteins with a light absorbing pigment attached called a chromophore. The chromophore is a linear tetrapyrrole called phytochromobilin.

There are two forms of phytochromes: red light absorbing, Pr, and far-red light absorbing, Pfr. Pfr, which is the active form of phytochromes, can be reverted to Pr, which is the inactive form, slowly by inducing darkness or more rapidly by irradiation by far-red light. The phytochrome apoprotein, a protein that together with a prosthetic group forms a particular biochemical molecule such as a hormone or enzyme, is synthesized in the Pr form. Upon binding the chromophore, the holoprotein, an apoprotein combined with its prosthetic group, becomes sensitive to light. If it absorbs red light it will change conformation to the biologically active Pfr form. The Pfr form can absorb far red light and switch back to the Pr form. The Pfr promotes and regulates photomorphogenesis in response to FR light, whereas Pr regulates de-etiolation in response to R light.

Most plants have multiple phytochromes encoded by different genes. The different forms of phytochrome control different responses but there is also redundancy so that in the absence of one phytochrome, another may take on the missing functions. There are five genes that encode phytochromes in the *Arabidopsis thaliana* genetic model, *PHYA-PHYE*. PHYA is involved in the regulation of photomorphogenesis in response to far-red light. PHYB is involved in regulating photoreversible seed germination in response to red light. PHYC mediates the response between PHYA and PHYB. PHYD and PHYE mediate elongation of the internode and control the time in which the plant flowers.

Molecular analyses of phytochrome and phytochrome-like genes in higher plants (ferns, mosses, algae) and photosynthetic bacteria have shown that phytochromes evolved from prokaryotic photoreceptors that predated the origin of plants.

Takuma Tanada observed that the root tips of barley adhered to the sides of a beaker with a negatively charged surface after being treated with red light, yet released after being exposed to far-red light. For mung bean it was the opposite, where far-red light exposure caused the root tips to adhere, and red light caused the roots to detach. This effect of red and far-red light on root tips is now known as the Tanada effect.

Blue Light

Plants contain multiple blue light photoreceptors which have different functions. Based on studies with action spectra, mutants and molecular analyses, it has been determined that higher plants contain at least 4, and probably 5, different blue light photoreceptors.

Cryptochromes were the first blue light receptors to be isolated and characterized from any organism, and are responsible for the blue light reactions in photomorphogenesis. The proteins use a flavin as a chromophore. The cryptochromes have evolved from microbial DNA-photolyase, an enzyme that carries out light-dependent repair of UV damaged DNA. Two cryptochromes have been identified in plants. There are two different forms of crytochromes, CRY1 and CRY2, which regulate the inhibition of hypocotyl elongation in response to blue light. Cryptochromes control stem elongation, leaf expansion, circadian rhythms and flowering time. In addition to blue light, cryptochromes also perceive long wavelength UV irradiation (UV-A). Since the cryptochromes were discovered in plants, several labs have identified homologous genes and photoreceptors in a number of other organisms, including humans, mice and flies.

There are blue light photoreceptors that are not a part of photomorphogenesis. For example, phototropin is the blue light photoreceptor that controls phototropism.

UV Light

Plants show various responses to UV light. UVR8 has been shown to be a UV-B receptor.

Photoperiodism

Photoperiodism is the physiological reaction of organisms to the length of day or night. It occurs in plants and animals. Photoperiodism can also be defined as the developmental responses of plants to the relative lengths of light and dark periods.

Plants

Many flowering plants (angiosperms) use a photoreceptor protein, such as phytochrome or cryptochrome, to sense seasonal changes in night length, or photoperiod, which they take as signals to flower. In a further subdivision, *obligate* photoperiodic plants absolutely require a long or short enough night before flowering, whereas *facultative* photoperiodic plants are more likely to flower under one condition.

Phytochrome comes in two forms: pr and pfr. Red light (which is present during the day) converts phytochrome to it's active form (pfr). This then triggers the plant to grow. In turn, far-red light is present in the shade or in the dark and this converts phytochrome from pfr to pr. Pr is the inactive form of phytochrome and will not allow for plant growth. This system of pfr to pr conversion allows the plant to sense when it is night and when it is day. Pfr can also be convereted back to Pr by a process known as dark reversion, where long periods of darkness trigger the conversion of Pfr. This is important in regards to plant flowering. Experiments by Halliday et. al showed that manipulations of the red-to far-red ratio in Arabidopsis can alter flowering. They discovered that plants tend to flower later when exposed to more red light, proving that red light is inhibitory to flowering. Other experiments have proven this by exposing plants to extra red-light in the middle of the night. A short-day plant will not flower if light is turned on for a few minutes in the middle of the night and a long-day plant can flower if exposed to more red-light in the middle of the night.

Cryptochromes are another type of photoreceptor that is important in photoperiodism. Cryptochromes absorb blue light and UV-A. Cryptochromes entrain the circadian clock to light. It has been found that both cryptochrome and phytochrome abundance relies on light and the amount of cryptochrome can change depending on day-length. This shows how important both of the photoreceptors are in regards to determining day-length.

In 1920, W. W. Garner and H. A. Allard published their discoveries on photoperiodism and felt it was the length of daylight that was critical, but it was later discovered that the length of the night was the controlling factor. Photoperiodic flowering plants are classified as *long-day plants* or *short-day plants* even though night is the critical factor because of the initial misunderstanding about daylight being the controlling factor. Along with long-day plants and short-day plants, there are plants that fall into a "dual-day length category". These plants are either long-short-day plants (LSDP) or short-long-day plants (SLDP). LSDPs flower after a series of long days followed by short days whereas SLDPs flower after a series of short days followed by long days. Each plant has a different length critical photoperiod, or critical night length.

Modern biologists believe that it is the coincidence of the active forms of phytochrome or cryptochrome, created by light during the daytime, with the rhythms of the circadian clock that allows plants to measure the length of the night. Other than flowering, photoperiodism in plants includes the growth of stems or roots during certain seasons and the loss of leaves. Artificial lighting can be used to induce extra-long days.

Long-day Plants

Long-day plants flower when the night length falls below their critical photoperiod. These plants typically flower in the northern hemisphere during late spring or early summer as days are getting longer. In the northern hemisphere, the longest day of the year (summer solstice) is on or about 21 June. After that date, days grow shorter (i.e. nights grow longer) until 21 December (the winter solstice). This situation is reversed in the southern hemisphere (i.e., longest day is 21 December and shortest day is 21 June).

Some long-day obligate plants are:

- Carnation (*Dianthus*)
- Henbane (*Hyoscyamus*)
- Oat (*Avena*)

Some long-day facultative plants are:

- Pea (*Pisum sativum*)
- Barley (*Hordeum vulgare*)
- Lettuce (*Lactuca sativa*)
- Wheat (*Triticum aestivum*)

Short-day Plants

Short-day plants flower when the night lengths exceed their critical photoperiod. They cannot flower under short nights or if a pulse of artificial light is shone on the plant for several minutes during the night; they require a continuous period of darkness before floral development can begin. Natural nighttime light, such as moonlight or lightning, is not of sufficient brightness or duration to interrupt flowering.

In general, short-day (i.e.long-night) plants flower as days grow shorter (and nights grow longer) after 21 June in the northern hemisphere, which is during summer or fall. The length of the dark period required to induce flowering differs among species and varieties of a species.

Photoperiodism affects flowering by inducing the shoot to produce floral buds instead of leaves and lateral buds.

Some short-day facultative plants are:

- Marijuana (*Cannabis*)
- Cotton (*Gossypium*)
- Rice (*Oryza*)
- Jowar (*Sorghum bicolor*)
- Green Gram (Mung bean, *Vigna radiata*)
- Soybeans (*Glycine max*)

Day-neutral Plants

Day-neutral plants, such as cucumbers, roses, and tomatoes, do not initiate flowering based on photoperiodism. Instead, they may initiate flowering after attaining a certain overall developmental stage or age, or in response to alternative environmental stimuli, such as vernalisation (a period of low temperature).

Animal

Daylength, and thus knowledge of the season of the year, is vital to many animals. A number of biological and behavioural changes are dependent on this knowledge. Together with temperature changes, photoperiod provokes changes in the color of fur and feathers, migration, entry into hibernation, sexual behaviour, and even the resizing of sexual organs.

The singing frequency of birds such as the canary depends on the photoperiod. In the spring, when the photoperiod increases (more daylight), the male canary's testes grow. As the testes grow, more androgens are secreted and song frequency increases. During autumn, when the photoperiod decreases (less daylight), the male canary's testes regress and androgen levels drop dramatically, resulting in decreased singing frequency. Not only is singing frequency dependent on the photoperiod but the song repertoire is also. The long photoperiod of spring results in a greater song repertoire. Autumn's shorter photoperiod results in a reduction in song repertoire.

These behavioral photoperiod changes in male canaries are caused by changes in the song center of the brain. As the photoperiod increases, the high vocal center (HVC) and the robust nucleus of the archistriatum (RA) increase in size. When the photoperiod decreases, these areas of the brain regress.

In mammals, daylength is registered in the suprachiasmatic nucleus (SCN), which is informed by retinal light-sensitive ganglion cells, which are not involved in vision. The information travels through the retinohypothalamic tract (RHT). Some mammals are highly seasonal, while humans' seasonality is largely believed to be evolutionary baggage.

Plant Pathology

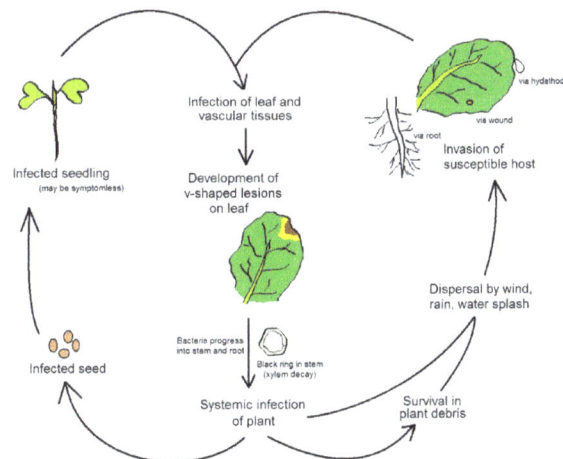

Life cycle of the black rot pathogen, *Xanthomonas campestris* pathovar *campes*

Plant pathology (also phytopathology) is the scientific study of diseases in plants caused by pathogens (infectious organisms) and environmental conditions (physiological factors). Organisms that cause infectious disease include fungi, oomycetes, bacteria, viruses, viroids, virus-like organisms, phytoplasmas, protozoa, nematodes and parasitic plants. Not included are ectoparasites like insects, mites, vertebrate, or other pests that affect plant health by consumption of plant tissues. Plant pathology also involves the study of pathogen identification, disease etiology, disease cycles, economic impact, plant disease epidemiology, plant disease resistance, how plant diseases affect humans and animals, pathosystem genetics, and management of plant diseases.

Overview

Control of plant diseases is crucial to the reliable production of food, and it provides significant reductions in agricultural use of land, water, fuel and other inputs. Plants in both natural and cultivated populations carry inherent disease resistance, but there are numerous examples of devastating plant disease impacts, as well as recurrent severe plant diseases. However, disease control is reasonably successful for most crops. Disease control is achieved by use of plants that have been bred for good resistance to many diseases, and by plant cultivation approaches such as crop rotation, use of pathogen-free seed, appropriate planting date and plant density, control of field moisture, and pesticide use. Across large regions and many crop species, it is estimated that diseases typically reduce plant yields by 10% every year in more developed settings, but yield loss to diseases often exceeds 20% in less developed settings. Continuing advances in the science of plant pathology are needed to improve disease control, and to keep up with changes in disease pressure caused by the ongoing evolution and movement of plant pathogens and by changes in agricultural practices. Plant diseases cause major economic losses for farmers worldwide. The Food and Agriculture Organization estimates indeed that pests and diseases are responsible for about 25% of crop loss. To solve this issue, new methods are needed to detect diseases and pests early, such as novel sensors that detect plant odours and spectroscopy and biophotonics that are able to diagnostic plant health and metabolism.

Plant Pathogens

Fungi

Most phytopathogenic fungi belong to the Ascomycetes and the Basidiomycetes.

The fungi reproduce both sexually and asexually via the production of spores and other structures. Spores may be spread long distances by air or water, or they may be soilborne. Many soil inhabiting fungi are capable of living saprotrophically, carrying out the part of their life cycle in the soil. These are known as facultative saprotrophs.

Fungal diseases may be controlled through the use of fungicides and other agriculture practices. However, new races of fungi often evolve that are resistant to various fungicides.

Biotrophic fungal pathogens colonize living plant tissue and obtain nutrients from living host cells. Necrotrophic fungal pathogens infect and kill host tissue and extract nutrients from the dead host cells. The powdery mildew and rice blast images, below. Significant fungal plant pathogens include:

Rice blast, caused by a necrotrophic fungus

Ascomycetes

- *Fusarium* spp. (causal agents of Fusarium wilt disease)

- *Thielaviopsis* spp. (causal agents of: canker rot, black root rot, *Thielaviopsis* root rot)

- *Verticillium* spp.

- *Magnaporthe grisea* (causal agent of rice blast)

- *Sclerotinia sclerotiorum* (causal agent of cottony rot)

Basidiomycetes

- *Ustilago* spp. (causal agents of smut)

- *Rhizoctonia* spp.

- *Phakospora pachyrhizi* (causal agent of soybean rust)

- *Puccinia* spp. (causal agents of severe rusts of virtually all cereal grains and cultivated grasses)

- *Armillaria* spp. (the so-called honey fungus species, which are virulent pathogens of trees and produce edible mushrooms)

Fungus-like Organisms

Oomycetes

The oomycetes are not true fungi but are fungus-like organisms. They include some of the most destructive plant pathogens including the genus *Phytophthora*, which includes the causal agents of potato late blight and sudden oak death. Particular species of oomycetes are responsible for root rot.

Despite not being closely related to the fungi, the oomycetes have developed very similar infection strategies. Oomycetes are capable of using effector proteins to turn off a plant's defenses in its infection process. Plant pathologists commonly group them with fungal pathogens.

Significant oomycete plant pathogens

- *Pythium* spp.

- *Phytophthora* spp., including the causal agent of the Great Irish Famine (1845–1849)

Phytomyxea

Some slime molds in Phytomyxea cause important diseases, including club root in cabbage and its relatives and powdery scab in potatoes. These are caused by species of *Plasmodiophora* and *Spongospora*, respectively.

Bacteria

Most bacteria that are associated with plants are actually saprotrophic and do no harm to the plant itself. However, a small number, around 100 known species, are able to cause disease. Bacterial diseases are much more prevalent in subtropical and tropical regions of the world.

Crown gall disease caused by Agrobacterium

Most plant pathogenic bacteria are rod-shaped (bacilli). In order to be able to colonize the plant they have specific pathogenicity factors. Five main types of bacterial pathogenicity factors are known: uses of cell wall–degrading enzymes, toxins, effector proteins, phytohormones and exopolysaccharides.

Pathogens such as *Erwinia* species use cell wall–degrading enzymes to cause soft rot. *Agrobacterium* species change the level of auxins to cause tumours with phytohormones. Exopolysaccharides are produced by bacteria and block xylem vessels, often leading to the death of the plant.

Bacteria control the production of pathogenicity factors via quorum sensing.

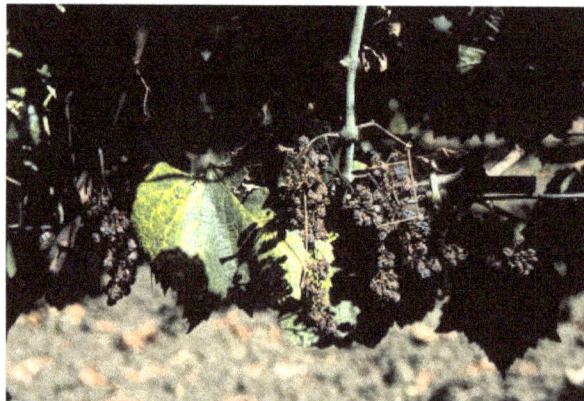

Vitis vinifera with "Ca. Phytoplasma vitis" infection

Significant bacterial plant pathogens:

- Burkholderia

- Proteobacteria

    o *Xanthomonas* spp.

    o *Pseudomonas* spp.

- Pseudomonas syringae pv. tomato causes tomato plants to produce less fruit, and it "continues to adapt to the tomato by minimizing its recognition by the tomato immune system."

Phytoplasmas ('Mycoplasma-like organisms') and Spiroplasmas

Phytoplasma and *Spiroplasma* are a genre of bacteria that lack cell walls and are related to the mycoplasmas, which are human pathogens. Together they are referred to as the mollicutes. They also tend to have smaller genomes than most other bacteria. They are normally transmitted by sap-sucking insects, being transferred into the plants phloem where it reproduces.

Tobacco mosaic virus

Viruses, Viroids and Virus-like Organisms

There are many types of plant virus, and some are even asymptomatic. Under normal circumstances, plant viruses cause only a loss of crop yield. Therefore, it is not economically viable to try to control them, the exception being when they infect perennial species, such as fruit trees.

Most plant viruses have small, single-stranded RNA genomes. However some plant viruses also have double stranded RNA or single or double stranded DNA genomes. These genomes may encode only three or four proteins: a replicase, a coat protein, a movement protein, in order to allow cell to cell movement through plasmodesmata, and sometimes a protein that allows transmission by a vector. Plant viruses can have several more proteins and employ many different molecular translation methods.

Plant viruses are generally transmitted from plant to plant by a vector, but mechanical and seed transmission also occur. Vector transmission is often by an insect (for example, aphids), but some fungi, nematodes, and protozoa have been shown to be viral vectors. In many cases, the insect and virus are specific for virus transmission such as the beet leafhopper that transmits the curly top virus causing disease in several crop plants.

Nematodes

Nematodes are small, multicellular wormlike animals. Many live freely in the soil, but there are some species that parasitize plant roots. They are a problem in tropical and subtropical regions of the world, where they may infect crops. Potato cyst nematodes (*Globodera pallida* and *G. rostochiensis*) are widely distributed in Europe and North and South America and cause $300 million worth of damage in Europe every year. Root knot nematodes have quite a large host range, they parasitize plant root systems and thus directly affect the uptake of water and nutrients needed for normal plant growth and reproduction, whereas cyst nematodes tend to be able to infect only a few species. Nematodes are able to cause radical changes in root cells in order to facilitate their lifestyle.

Root-knot nematode galls

Protozoa and Algae

There are a few examples of plant diseases caused by protozoa (e.g., *Phytomonas*, a kinetoplastid). They are transmitted as zoospores that are very durable, and may be able to survive in a resting state in the soil for many years. They have also been shown to transmit plant viruses.

When the motile zoospores come into contact with a root hair they produce a plasmodium and invade the roots.

Some colourless parasitic algae (e.g., *Cephaleuros*) also cause plant diseases.

Parasitic Plants

Parasitic plants such as mistletoe and dodder are included in the study of phytopathology. Dodder, for example, is used as a conduit either for the transmission of viruses or virus-like agents from a host plant to a plant that is not typically a host or for an agent that is not graft-transmissible.

Common Pathogenic Infection Methods

* Cell wall-degrading enzymes: These are used to break down the plant cell wall in order to release the nutrients inside.

* Toxins: These can be non-host-specific, which damage all plants, or host-specific, which cause damage only on a host plant.

* Effector proteins: These can be secreted into the extracellular environment or directly into the host cell, often via the Type three secretion system. Some effectors are known to suppress host defense processes. This can include: reducing the plants internal signaling mechanisms or reduction of phytochemicals production. Bacteria, fungus and oomycetes are known for this function.

Physiological Plant Disorders

Abiotic disorders can be caused by natural processes such as drought, frost, snow and hail; flooding and poor drainage; nutrient deficiency; deposition of mineral salts such as sodium chloride and gypsum; windburn and breakage by storms; wildfires. Similar disorders (usually classed as abiotic) can be caused by human intervention, resulting in soil compaction, pollution of air and soil, salinisation caused by irrigation and road salting, over-application of herbicides, clumsy handling (e.g. lawnmower damage to trees), and vandalism.

Orchid leaves with viral infections

Management

Quarantine

A diseased patch of vegetation or individual plants can be isolated from other, healthy growth. Specimens may be destroyed or relocated into a greenhouse for treatment or study. Another option is to avoid the introduction of harmful nonnative organisms by controlling all human traffic and activity (e.g., AQIS), although legislation and enforcement are crucial in order to ensure lasting effectiveness.

Cultural

Farming in some societies is kept on a small scale, tended by peoples whose culture includes farming traditions going back to ancient times. (An example of such traditions would be lifelong training in techniques of plot terracing, weather anticipation and response, fertilization, grafting, seed care, and dedicated gardening.) Plants that are intently monitored often benefit from not only active external protection but also a greater overall vigor. While primitive in the sense of being the most labor-intensive solution by far, where practical or necessary it is more than adequate.

Plant Resistance

Sophisticated agricultural developments now allow growers to choose from among systematically cross-bred species to ensure the greatest hardiness in their crops, as suited for a particular region's pathological profile. Breeding practices have been perfected over centuries, but with the advent of genetic manipulation even finer control of a crop's immunity traits is possible. The engineering of food plants may be less rewarding, however, as higher output is frequently offset by popular suspicion and negative opinion about this "tampering" with nature.

Chemical

Many natural and synthetic compounds can be employed to combat the above threats. This method works by directly eliminating disease-causing organisms or curbing their spread; however, it has been shown to have too broad an effect, typically, to be good for the local ecosystem. From an economic standpoint, all but the simplest natural additives may disqualify a product from "organic" status, potentially reducing the value of the yield.

Biological

Crop rotation may be an effective means to prevent a parasitic population from becoming well-established, as an organism affecting leaves would be starved when the leafy crop is replaced by a tuberous type, etc. Other means to undermine parasites without attacking them directly may exist.

Integrated

The use of two or more of these methods in combination offers a higher chance of effectiveness.

References

- Parent, S.-E. et al. 2013. The plant ionome revisited by the nutrient balance concept. Front. Plant Sci. 4. doi:10.3389/fpls.2013.00039

- Fosket, Donald E. (1994). Plant Growth and Development: A Molecular Approach. San Diego: Academic Press. pp. 498–509. ISBN 0-12-262430-0

- "AAPFCO Board of Directors 2006 Mid-Year Meeting" (PDF). Association of American Plant Food Control Officials. Retrieved 18 July 2011

- Tarakhovskaya, E. R.; Maslov, Yu; Shishova, M. F. (2007). "Phytohormones in algae". Russian Journal of Plant Physiology. 54 (2): 163–170. doi:10.1134/s1021443707020021

- Bittsanszky, A. et al. 2015. Overcoming ammonium toxicity. Plant Sci. Int. J. Exp. Plant Biol. 231C, 184–190. doi:10.1016/j.plantsci.2014.12.005

- Marschner, Petra, ed. (2012). Marschner's mineral nutrition of higher plants (3rd ed.). Amsterdam: Elsevier/ Academic Press. ISBN 9780123849052

- Nicole Davis (September 9, 2009). "Genome of Irish potato famine pathogen decoded". Haas et al. Broad Institute of MIT and Harvard. Retrieved 24 July 2012

- Kermode AR (December 2005). "Role of Abscisic Acid in Seed Dormancy". J Plant Growth Regul. 24 (4): 319–344. doi:10.1007/s00344-005-0110-2

- Öpik, Helgi; Rolfe, Stephen A.; Willis, Arthur John; Street, Herbert Edward (2005). The physiology of flowering plants (4th ed.). Cambridge University Press. p. 191. ISBN 978-0-521-66251-2

- "Scientists discover how deadly fungal microbes enter host cells". (VBI) at Virginia Tech affiliates. Physorg. July 22, 2010. Retrieved July 31, 2012

- Casal, J.J. (2014). "Light perception and signalling by phytochrome A". Journal of Experimental Botany. 65 (11).: 2835–2845

- Sipes DL, Einset JW (August 1983). "Cytokinin stimulation of abscission in lemon pistil explants". J Plant Growth Regul. 2 (1–3): 73–80. doi:10.1007/BF02042235

- Osborne, Daphné J.; McManus, Michael T. (2005). Hormones, signals and target cells in plant development. Cambridge University Press. p. 158. ISBN 978-0-521-33076-3

- "Plasmopara viticola, the Cause of Downy Mildew of Grapes". The Origin of Plant Pathology and The Potato Famine, and Other Stories of Plant Diseases. Retrieved 4 February 2015

- Ulm, Roman; Jenkins, Gareth I (2015-06-30). "Q&A: How do plants sense and respond to UV-B radiation?". BMC Biology. 13 (1). PMC 4484705. PMID 26123292. doi:10.1186/s12915-015-0156-y

- Jackson RW (editor). (2009). Plant Pathogenic Bacteria: Genomics and Molecular Biology. Caister Academic Press. ISBN 978-1-904455-37-0

- Miranda, Stephen R.; Barker, Bruce (August 4, 2009). "Silicon: Summary of Extraction Methods". Harsco Minerals. Retrieved 18 July 2011

- Creamer, Rebecca; H. Hubble; A. Lewis (May 2005). "Curtovirus Infection of Chile Pepper in New Mexico". Plant Diseases. 89 (5): 480–486. doi:10.1094/PD-89-0480

Essential Aspects of Plant Science

Photosynthesis is the process that is responsible for providing nutrition to plants as well as producing oxygen. The other important aspects related to plant science are phytochemistry, seedling, tropism and flora. This chapter discusses the essential aspects of plant science in a critical manner providing key analysis to the subject matter.

Photosynthesis

Photosynthesis is a process used by plants and other organisms to convert light energy into chemical energy that can later be released to fuel the organisms' activities (energy transformation). This chemical energy is stored in carbohydrate molecules, such as sugars, which are synthesized from carbon dioxide and water – hence the name *photosynthesis*, from the Greek *phōs*, "light", and, *synthesis*, "putting together". In most cases, oxygen is also released as a waste product. Most plants, most algae, and cyanobacteria perform photosynthesis; such organisms are called photoautotrophs. Photosynthesis is largely responsible for producing and maintaining the oxygen content of the Earth's atmosphere, and supplies all of the organic compounds and most of the energy necessary for life on Earth.

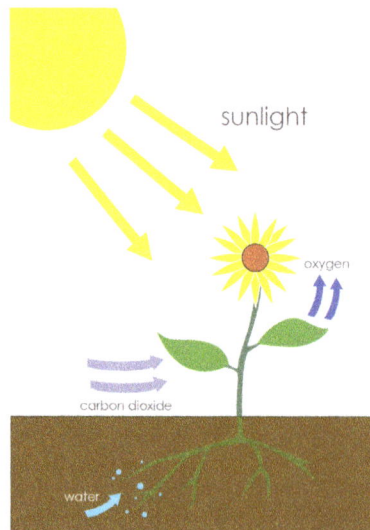

Schematic of photosynthesis in plants. The carbohydrates produced are stored in or used by the plant.

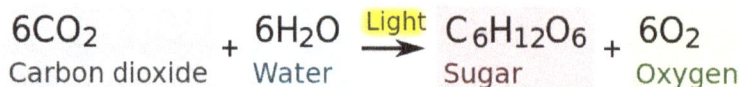

$$6CO_2 + 6H_2O \xrightarrow{\text{Light}} C_6H_{12}O_6 + 6O_2$$

Carbon dioxide   Water   Sugar   Oxygen

Overall equation for the type of photosynthesis that occurs in plants

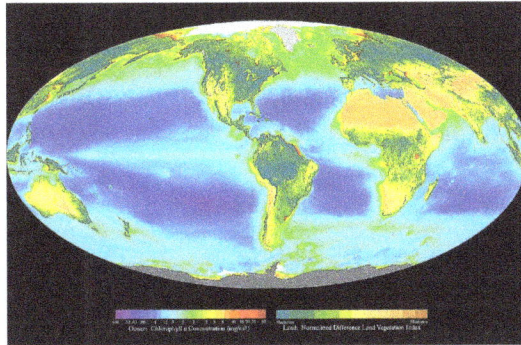

Composite image showing the global distribution of photosynthesis, including both oceanic phytoplankton and terrestrial vegetation. Dark red and blue-green indicate regions of high photosynthetic activity in the ocean and on land, respectively.

Although photosynthesis is performed differently by different species, the process always begins when energy from light is absorbed by proteins called reaction centres that contain green chlorophyll pigments. In plants, these proteins are held inside organelles called chloroplasts, which are most abundant in leaf cells, while in bacteria they are embedded in the plasma membrane. In these light-dependent reactions, some energy is used to strip electrons from suitable substances, such as water, producing oxygen gas. The hydrogen freed by the splitting of water is used in the creation of two further compounds that act as an immediate energy storage means: reduced nicotinamide adenine dinucleotide phosphate (NADPH) and adenosine triphosphate (ATP), the "energy currency" of cells.

In plants, algae and cyanobacteria, long-term energy storage in the form of sugars is produced by a subsequent sequence of light-independent reactions called the Calvin cycle; some bacteria use different mechanisms, such as the reverse Krebs cycle, to achieve the same end. In the Calvin cycle, atmospheric carbon dioxide is incorporated into already existing organic carbon compounds, such as ribulose bisphosphate (RuBP). Using the ATP and NADPH produced by the light-dependent reactions, the resulting compounds are then reduced and removed to form further carbohydrates, such as glucose.

The first photosynthetic organisms probably evolved early in the evolutionary history of life and most likely used reducing agents such as hydrogen or hydrogen sulfide, rather than water, as sources of electrons. Cyanobacteria appeared later; the excess oxygen they produced contributed directly to the oxygenation of the Earth, which rendered the evolution of complex life possible. Today, the average rate of energy capture by photosynthesis globally is approximately 130 terawatts, which is about three times the current power consumption of human civilization. Photosynthetic organisms also convert around 100–115 thousand million metric tonnes of carbon into biomass per year.

Overview

Photosynthetic organisms are photoautotrophs, which means that they are able to synthesize food directly from carbon dioxide and water using energy from light. However, not all organisms that use light as a source of energy carry out photosynthesis; photoheterotrophs use organic compounds, rather than carbon dioxide, as a source of carbon. In plants, algae, and cyanobacteria, photosynthesis releases oxygen. This is called *oxygenic photosynthesis* and is by far the most common type of photosynthesis used by living organisms. Although there are some differences between oxygenic photosynthesis in plants, algae, and cyanobacteria, the overall process is quite similar in these organisms. There are also many varieties of anoxygenic photosynthesis, used mostly by certain types of bacteria, which consume carbon dioxide but do not release oxygen.

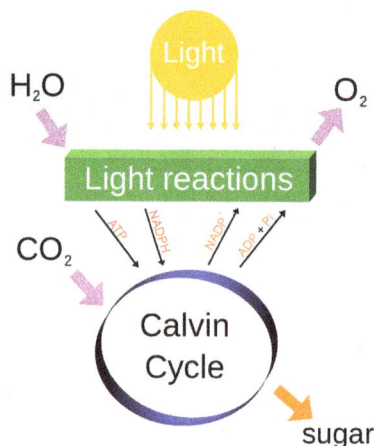

Photosynthesis changes sunlight into chemical energy, splits water to liberate O_2, and fixes CO_2 into sugar.

Carbon dioxide is converted into sugars in a process called carbon fixation. Photosynthesis provides the energy in the form of free electrons that are used to split carbon from carbon dioxide that is then used to fix that carbon once again as carbohydrate. Carbon fixation is an endothermic redox reaction, so photosynthesis supplies the energy that drives both process. In general outline, photosynthesis is the opposite of cellular respiration, in which glucose and other compounds are oxidized to produce carbon dioxide and water, and to release chemical energy (an exothermic reaction) to drive the organism's metabolism. The two processes, reduction of carbon dioxide to carbohydrate and then later oxidation of the carbohydrate, are distinct: photosynthesis and cellular respiration take place through a different sequence of chemical reactions and in different cellular compartments.

The general equation for photosynthesis as first proposed by Cornelius van Niel is therefore:

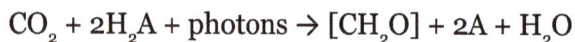

$$CO_2 + 2H_2A + photons \rightarrow [CH_2O] + 2A + H_2O$$

carbon dioxide + electron donor + light energy $\rightarrow$ carbohydrate + oxidized electron donor + water

Since water is used as the electron donor in oxygenic photosynthesis, the equation for this process is:

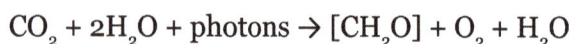

$$CO_2 + 2H_2O + photons \rightarrow [CH_2O] + O_2 + H_2O$$

carbon dioxide + water + light energy $\rightarrow$ carbohydrate + oxygen + water

This equation emphasizes that water is both a reactant in the light-dependent reaction and a product of the light-independent reaction, but canceling n water molecules from each side gives the net equation:

$$CO_2 + H_2O + photons \rightarrow [CH_2O] + O_2$$

carbon dioxide + water + light energy $\rightarrow$ carbohydrate + oxygen

Other processes substitute other compounds (such as arsenite) for water in the electron-supply role; for example some microbes use sunlight to oxidize arsenite to arsenate: The equation for this reaction is:

$$CO_2 + (AsO_3^{3-}) + photons \rightarrow (AsO_4^{3-}) + CO$$

carbon dioxide + arsenite + light energy $\rightarrow$ arsenate + carbon monoxide (used to build other compounds in subsequent reactions)

Photosynthesis occurs in two stages. In the first stage, *light-dependent reactions* or *light reactions* capture the energy of light and use it to make the energy-storage molecules ATP and NADPH. During the second stage, the *light-independent reactions* use these products to capture and reduce carbon dioxide.

Most organisms that utilize oxygenic photosynthesis use visible light for the light-dependent reactions, although at least three use shortwave infrared or, more specifically, far-red radiation.

Some organisms employ even more radical variants of photosynthesis. Some archea use a simpler method that employs a pigment similar to those used for vision in animals. The bacteriorhodopsin changes its configuration in response to sunlight, acting as a proton pump. This produces a proton gradient more directly, which is then converted to chemical energy. The process does not involve carbon dioxide fixation and does not release oxygen, and seems to have evolved separately from the more common types of photosynthesis.

Photosynthetic Membranes and Organelles

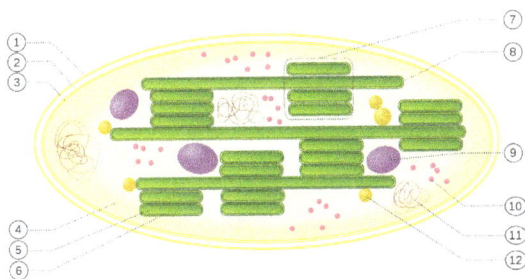

Chloroplast ultrastructure:
1. outer membrane
2. intermembrane space
3. inner membrane (1+2+3: envelope)
4. stroma (aqueous fluid)
5. thylakoid lumen (inside of thylakoid)
6. thylakoid membrane
7. granum (stack of thylakoids)
8. thylakoid (lamella)
9. starch
10. ribosome
11. plastidial DNA
12. plastoglobule (drop of lipids)

In photosynthetic bacteria, the proteins that gather light for photosynthesis are embedded in cell membranes. In its simplest form, this involves the membrane surrounding the cell itself. However, the membrane may be tightly folded into cylindrical sheets called thylakoids, or bunched up into round vesicles called *intracytoplasmic membranes*. These structures can fill most of the interior of a cell, giving the membrane a very large surface area and therefore increasing the amount of light that the bacteria can absorb.

In plants and algae, photosynthesis takes place in organelles called chloroplasts. A typical plant cell contains about 10 to 100 chloroplasts. The chloroplast is enclosed by a membrane. This membrane is composed of a phospholipid inner membrane, a phospholipid outer membrane, and an intermembrane space. Enclosed by the membrane is an aqueous fluid called the stroma. Embedded within the stroma are stacks of thylakoids (grana), which are the site of photosynthesis. The thylakoids appear as flattened disks. The thylakoid itself is enclosed by the thylakoid membrane, and within the enclosed volume is a lumen or thylakoid space. Embedded in the thylakoid membrane are integral and peripheral membrane protein complexes of the photosynthetic system.

Plants absorb light primarily using the pigment chlorophyll. The green part of the light spectrum is not absorbed but is reflected which is the reason that most plants have a green color. Besides chlorophyll, plants also use pigments such as carotenes and xanthophylls. Algae also use chlorophyll, but various other pigments are present, such as phycocyanin, carotenes, and xanthophylls in green algae, phycoerythrin in red algae (rhodophytes) and fucoxanthin in brown algae and diatoms resulting in a wide variety of colors.

These pigments are embedded in plants and algae in complexes called antenna proteins. In such proteins, the pigments are arranged to work together. Such a combination of proteins is also called a light-harvesting complex.

Although all cells in the green parts of a plant have chloroplasts, the majority of those are found in specially adapted structures called leaves. Certain species adapted to conditions of strong sunlight and aridity, such as many Euphorbia and cactus species, have their main photosynthetic organs in their stems. The cells in the interior tissues of a leaf, called the mesophyll, can contain between 450,000 and 800,000 chloroplasts for every square millimeter of leaf. The surface of the leaf is coated with a water-resistant waxy cuticle that protects the leaf from excessive evaporation of water and decreases the absorption of ultraviolet or blue light to reduce heating. The transparent epidermis layer allows light to pass through to the palisade mesophyll cells where most of the photosynthesis takes place.

Light-dependent Reactions

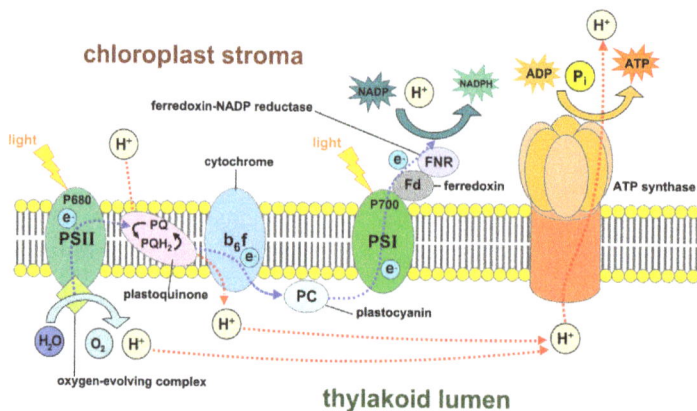

Light-dependent reactions of photosynthesis at the thylakoid membrane

In the light-dependent reactions, one molecule of the pigment chlorophyll absorbs one photon and loses one electron. This electron is passed to a modified form of chlorophyll called pheophytin, which passes the electron to a quinone molecule, starting the flow of electrons down an electron transport chain that leads to the ultimate reduction of NADP to NADPH. In addition, this creates a proton gradient (energy gradient) across the chloroplast membrane, which is used by ATP synthase in the synthesis of ATP. The chlorophyll molecule ultimately regains the electron it lost when a water molecule is split in a process called photolysis, which releases a dioxygen (O_2) molecule as a waste product.

The overall equation for the light-dependent reactions under the conditions of non-cyclic electron flow in green plants is:

$$2\,H_2O + 2\,NADP^+ + 3\,ADP + 3\,P_i + light \rightarrow 2\,NADPH + 2\,H^+ + 3\,ATP + O_2$$

Not all wavelengths of light can support photosynthesis. The photosynthetic action spectrum depends on the type of accessory pigments present. For example, in green plants, the action spectrum resembles the absorption spectrum for chlorophylls and carotenoids with absorption peaks in violet-blue and red light. In red algae, the action spectrum is blue-green light, which allows these algae to use the blue end of the spectrum to grow in the deeper waters that filter out the longer wavelengths (red light) used by above ground green plants. The non-absorbed part of the light spectrum is what gives photosynthetic organisms their color (e.g., green plants, red algae, purple bacteria) and is the least effective for photosynthesis in the respective organisms.

Z scheme

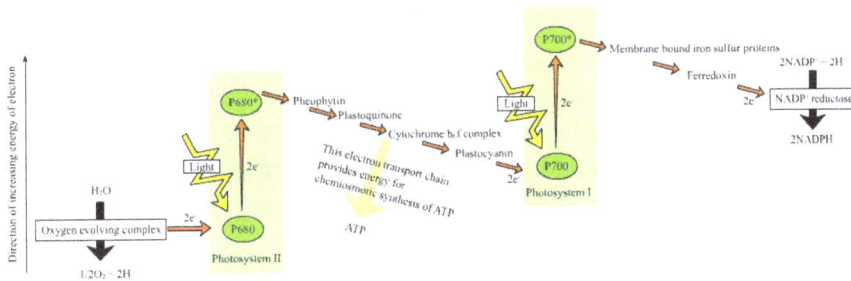

The "Z scheme"

In plants, light-dependent reactions occur in the thylakoid membranes of the chloroplasts where they drive the synthesis of ATP and NADPH. The light-dependent reactions are of two forms: cyclic and non-cyclic.

In the non-cyclic reaction, the photons are captured in the light-harvesting antenna complexes of photosystem II by chlorophyll and other accessory pigments. The absorption of a photon by the antenna complex frees an electron by a process called photoinduced charge separation. The antenna system is at the core of the chlorophyll molecule of the photosystem II reaction center. That freed electron is transferred to the primary electron-acceptor molecule, pheophytin. As the electrons are shuttled through an electron transport chain (the so-called Z-scheme shown in the diagram), it initially functions to generate a chemiosmotic potential by pumping proton cations (H^+) across the membrane and into the thylakoid space. An ATP synthase enzyme uses that chemiosmotic potential to make ATP during photophosphorylation, whereas NADPH is a product of the terminal redox reaction in the *Z-scheme*. The electron enters a chlorophyll molecule in

Photosystem I. There it is further excited by the light absorbed by that photosystem. The electron is then passed along a chain of electron acceptors to which it transfers some of its energy. The energy delivered to the electron acceptors is used to move hydrogen ions across the thylakoid membrane into the lumen. The electron is eventually used to reduce the co-enzyme NADP with a H⁺ to NADPH (which has functions in the light-independent reaction); at that point, the path of that electron ends.

The cyclic reaction is similar to that of the non-cyclic, but differs in that it generates only ATP, and no reduced NADP (NADPH) is created. The cyclic reaction takes place only at photosystem I. Once the electron is displaced from the photosystem, the electron is passed down the electron acceptor molecules and returns to photosystem I, from where it was emitted, hence the name *cyclic reaction*.

Water Photolysis

The NADPH is the main reducing agent produced by chloroplasts, which then goes on to provide a source of energetic electrons in other cellular reactions. Its production leaves chlorophyll in photosystem I with a deficit of electrons (chlorophyll has been oxidized), which must be balanced by some other reducing agent that will supply the missing electron. The excited electrons lost from chlorophyll from photosystem I are supplied from the electron transport chain by plastocyanin. However, since photosystem II is the first step of the *Z-scheme*, an external source of electrons is required to reduce its oxidized chlorophyll *a* molecules. The source of electrons in green-plant and cyanobacterial photosynthesis is water. Two water molecules are oxidized by four successive charge-separation reactions by photosystem II to yield a molecule of diatomic oxygen and four hydrogen ions; the electrons yielded are transferred to a redox-active tyrosine residue that then reduces the oxidized chlorophyll *a* (called P680) that serves as the primary light-driven electron donor in the photosystem II reaction center. That photo receptor is in effect reset and is then able to repeat the absorption of another photon and the release of another photo-dissociated electron. The oxidation of water is catalyzed in photosystem II by a redox-active structure that contains four manganese ions and a calcium ion; this oxygen-evolving complex binds two water molecules and contains the four oxidizing equivalents that are used to drive the water-oxidizing reaction. Photosystem II is the only known biological enzyme that carries out this oxidation of water. The hydrogen ions released contribute to the transmembrane chemiosmotic potential that leads to ATP synthesis. Oxygen is a waste product of light-dependent reactions, but the majority of organisms on Earth use oxygen for cellular respiration, including photosynthetic organisms.

Light-independent Reactions

Calvin Cycle

In the light-independent (or "dark") reactions, the enzyme RuBisCO captures CO_2 from the atmosphere and, in a process called the Calvin-Benson cycle, it uses the newly formed NADPH and releases three-carbon sugars, which are later combined to form sucrose and starch. The overall equation for the light-independent reactions in green plants is

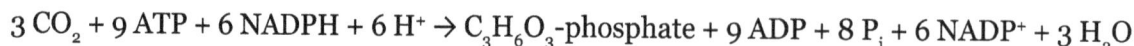

$$3\ CO_2 + 9\ ATP + 6\ NADPH + 6\ H^+ \rightarrow C_3H_6O_3\text{-phosphate} + 9\ ADP + 8\ P_i + 6\ NADP^+ + 3\ H_2O$$

Carbon fixation produces the intermediate three-carbon sugar product, which is then converted to the final carbohydrate products. The simple carbon sugars produced by photosynthesis are then used in the forming of other organic compounds, such as the building material cellulose, the precursors for lipid and amino acid biosynthesis, or as a fuel in cellular respiration. The latter occurs not only in plants but also in animals when the energy from plants is passed through a food chain.

The fixation or reduction of carbon dioxide is a process in which carbon dioxide combines with a five-carbon sugar, ribulose 1,5-bisphosphate, to yield two molecules of a three-carbon compound, glycerate 3-phosphate, also known as 3-phosphoglycerate. Glycerate 3-phosphate, in the presence of ATP and NADPH produced during the light-dependent stages, is reduced to glyceraldehyde 3-phosphate. This product is also referred to as 3-phosphoglyceraldehyde (PGAL) or, more generically, as triose phosphate. Most (5 out of 6 molecules) of the glyceraldehyde 3-phosphate produced is used to regenerate ribulose 1,5-bisphosphate so the process can continue. The triose phosphates not thus "recycled" often condense to form hexose phosphates, which ultimately yield sucrose, starch and cellulose. The sugars produced during carbon metabolism yield carbon skeletons that can be used for other metabolic reactions like the production of amino acids and lipids.

Carbon Concentrating Mechanisms

On land

In hot and dry conditions, plants close their stomata to prevent water loss. Under these conditions, CO_2 will decrease and oxygen gas, produced by the light reactions of photosynthesis, will increase, causing an increase of photorespiration by the oxygenase activity of ribulose-1,5-bisphosphate carboxylase/oxygenase and decrease in carbon fixation. Some plants have evolved mechanisms to increase the CO_2 concentration in the leaves under these conditions.

Plants that use the C_4 carbon fixation process chemically fix carbon dioxide in the cells of the mesophyll by adding it to the three-carbon molecule phosphoenolpyruvate (PEP), a reaction catalyzed by an enzyme called PEP carboxylase, creating the four-carbon organic acid oxaloacetic acid. Oxaloacetic acid or malate synthesized by this process is then translocated to specialized bundle sheath cells where the enzyme RuBisCO and other Calvin cycle enzymes are located, and where CO_2 released by decarboxylation of the four-carbon acids is then fixed by RuBisCO activity to the three-carbon 3-phosphoglyceric acids. The physical separation of RuBisCO from the oxygen-generating light reactions reduces photorespiration and increases CO_2 fixation and, thus, the photosynthetic capacity of the leaf. C_4 plants can produce more sugar than C_3 plants in conditions of high light and temperature. Many important crop plants are C_4 plants, including maize, sorghum, sugarcane, and millet. Plants that do not use PEP-carboxylase in carbon fixation are called C_3 plants because the primary carboxylation reaction, catalyzed by RuBisCO, produces the three-carbon 3-phosphoglyceric acids directly in the Calvin-Benson cycle. Over 90% of plants use C_3 carbon fixation, compared to 3% that use C_4 carbon fixation; however, the evolution of C_4 in over 60 plant lineages makes it a striking example of convergent evolution.

Xerophytes, such as cacti and most succulents, also use PEP carboxylase to capture carbon dioxide in a process called Crassulacean acid metabolism (CAM). In contrast to C_4 metabolism, which *spatially* separates the CO_2 fixation to PEP from the Calvin cycle, CAM *temporally* separates these two processes. CAM plants have a different leaf anatomy from C_3 plants, and fix the CO_2 at night, when their stomata are open. CAM plants store the CO_2 mostly in the form of malic acid via carboxylation of phosphoenolpyruvate to oxaloacetate, which is then reduced to malate. Decarboxylation of malate during the day releases CO_2 inside the leaves, thus allowing carbon fixation to 3-phosphoglycerate by RuBisCO. Sixteen thousand species of plants use CAM.

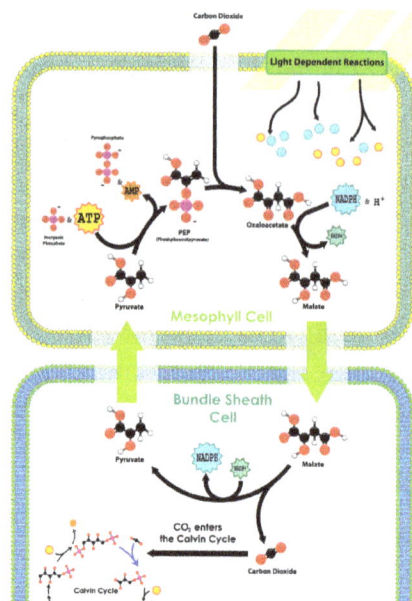

Overview of C4 carbon fixation

In Water

Cyanobacteria possess carboxysomes, which increase the concentration of CO_2 around RuBisCO to increase the rate of photosynthesis. An enzyme, carbonic anhydrase, located within the carboxysome releases CO_2 from the dissolved hydrocarbonate ions (HCO_3^-). Before the CO_2 diffuses out it is quickly sponged up by RuBisCO, which is concentrated within the carboxysomes. HCO_3^- ions are made from CO_2 outside the cell by another carbonic anhydrase and are actively pumped into the cell by a membrane protein. They cannot cross the membrane as they are charged, and within the cytosol they turn back into CO_2 very slowly without the help of carbonic anhydrase. This causes the HCO_3^- ions to accumulate within the cell from where they diffuse into the carboxysomes. Pyrenoids in algae and hornworts also act to concentrate CO_2 around rubisco.

Order and Kinetics

The overall process of photosynthesis takes place in four stages:

Stage	Description	Time scale
1	Energy transfer in antenna chlorophyll (thylakoid membranes)	femtosecond to picosecond
2	Transfer of electrons in photochemical reactions (thylakoid membranes)	picosecond to nanosecond
3	Electron transport chain and ATP synthesis (thylakoid membranes)	microsecond to millisecond
4	Carbon fixation and export of stable products	millisecond to second

Efficiency

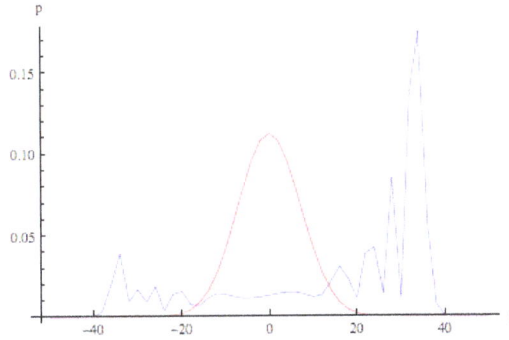

Probability distribution resulting from one-dimensional discrete time random walks. The quantum walk created using the Hadamard coin is plotted (blue) vs a classical walk (red) after 50 time steps.

Plants usually convert light into chemical energy with a photosynthetic efficiency of 3–6%. Absorbed light that is unconverted is dissipated primarily as heat, with a small fraction (1–2%) re-emitted as chlorophyll fluorescence at longer (redder) wavelengths. A fact that allows measurement of the light reaction of photosynthesis by using chlorophyll fluorometers.

Actual plants' photosynthetic efficiency varies with the frequency of the light being converted, light intensity, temperature and proportion of carbon dioxide in the atmosphere, and can vary from 0.1% to 8%. By comparison, solar panels convert light into electric energy at an efficiency of approximately 6–20% for mass-produced panels, and above 40% in laboratory devices.

The efficiency of both light and dark reactions can be measured but the relationship between the two can be complex. For example, the ATP and NADPH energy molecules, created by the light reaction, can be used for carbon fixation or for photorespiration in C_3 plants. Electrons may also flow to other electron sinks. For this reason, it is not uncommon for authors to differentiate between work done under non-photorespiratory conditions and under photorespiratory conditions.

Chlorophyll fluorescence of photosystem II can measure the light reaction, and Infrared gas analyzers can measure the dark reaction. It is also possible to investigate both at the same time using an integrated chlorophyll fluorometer and gas exchange system, or by using two separate systems together. Infrared gas analyzers and some moisture sensors are sensitive enough to measure the photosynthetic assimilation of CO_2, and of ΔH_2O using reliable methods CO_2 is commonly measured in $\mu mols/m^2/s^{-1}$, parts per million or volume per million and H_2O is commonly measured in $mmol/m^2/s^{-1}$ or in mbars. By measuring CO_2 assimilation, ΔH_2O, leaf temperature, barometric pressure, leaf area, and photosynthetically active radiation or PAR, it becomes possible to estimate, "A" or carbon assimilation, "E" or transpiration, "gs" or stomatal conductance, and Ci or intracellular CO_2. However, it is more common to used chlorophyll fluorescence for plant stress measurement, where appropriate, because the most commonly used measuring parameters FV/FM and Y(II) or F/FM' can be made in a few seconds, allowing the measurement of larger plant populations.

Gas exchange systems that offer control of CO_2 levels, above and below ambient, allow the common practice of measurement of A/Ci curves, at different CO_2 levels, to characterize a plant's photosynthetic response.

Integrated chlorophyll fluorometer – gas exchange systems allow a more precise measure of photosynthetic response and mechanisms. While standard gas exchange photosynthesis systems can measure Ci, or substomatal CO_2 levels, the addition of integrated chlorophyll fluorescence measurements allows a more precise measurement of C_C to replace Ci. The estimation of CO_2 at the site of carboxylation in the chloroplast, or C_C, becomes possible with the measurement of mesophyll conductance or g_m using an integrated system.

Photosynthesis measurement systems are not designed to directly measure the amount of light absorbed by the leaf. But analysis of chlorophyll-fluorescence, P700- and P515-absorbance and gas exchange measurements reveal detailed information about e.g. the photosystems, quantum efficiency and the CO_2 assimilation rates. With some instruments even wavelength-dependency of the photosynthetic efficiency can be analyzed.

A phenomenon known as quantum walk increases the efficiency of the energy transport of light significantly. In the photosynthetic cell of an algae, bacterium, or plant, there are light-sensitive molecules called chromophores arranged in an antenna-shaped structure named a photocomplex. When a photon is absorbed by a chromophore, it is converted into a quasiparticle referred to as an exciton, which jumps from chromophore to chromophore towards the reaction center of the photocomplex, a collection of molecules that traps its energy in a chemical form that makes it accessible for the cell's metabolism. The exciton's wave properties enable it to cover a wider area and try out several possible paths simultaneously, allowing it to instantaneously "choose" the most efficient route, where it will have the highest probability of arriving at its destination in the minimum possible time. Because that quantum walking takes place at temperatures far higher than quantum phenomena usually occur, it is only possible over very short distances, due to obstacles in the form of destructive interference that come into play. These obstacles cause the particle to lose its wave properties for an instant before it regains them once again after it is freed from its locked position through a classic "hop". The movement of the electron towards the photo center is therefore covered in a series of conventional hops and quantum walks.

Evolution

Early photosynthetic systems, such as those in green and purple sulfur and green and purple nonsulfur bacteria, are thought to have been anoxygenic, and used various other molecules as electron donors rather than water. Green and purple sulfur bacteria are thought to have used hydrogen and sulfur as electron donors. Green nonsulfur bacteria used various amino and other organic acids as an electron donor. Purple nonsulfur bacteria used a variety of nonspecific organic molecules. The use of these molecules is consistent with the geological evidence that Earth's early atmosphere was highly reducing at that time.

Fossils of what are thought to be filamentous photosynthetic organisms have been dated at 3.4 billion years old.

The main source of oxygen in the Earth's atmosphere derives from oxygenic photosynthesis, and its first appearance is sometimes referred to as the oxygen catastrophe. Geological evidence suggests that oxygenic photosynthesis, such as that in cyanobacteria, became important during the Paleoproterozoic era around 2 billion years ago. Modern photosynthesis in plants and most photosynthetic prokaryotes is oxygenic. Oxygenic photosynthesis uses water as an electron donor, which is oxidized to molecular oxygen (O_2) in the photosynthetic reaction center.

Symbiosis and the Origin of Chloroplasts

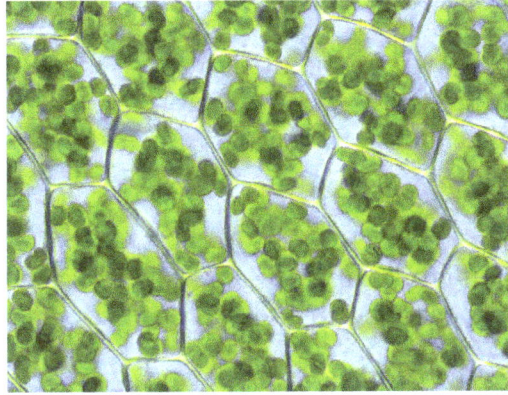

Plant cells with visible chloroplasts (from a moss, *Plagiomnium affine*)

Several groups of animals have formed symbiotic relationships with photosynthetic algae. These are most common in corals, sponges and sea anemones. It is presumed that this is due to the particularly simple body plans and large surface areas of these animals compared to their volumes. In addition, a few marine mollusks *Elysia viridis* and *Elysia chlorotica* also maintain a symbiotic relationship with chloroplasts they capture from the algae in their diet and then store in their bodies. This allows the mollusks to survive solely by photosynthesis for several months at a time. Some of the genes from the plant cell nucleus have even been transferred to the slugs, so that the chloroplasts can be supplied with proteins that they need to survive.

An even closer form of symbiosis may explain the origin of chloroplasts. Chloroplasts have many similarities with photosynthetic bacteria, including a circular chromosome, prokaryotic-type ribosome, and similar proteins in the photosynthetic reaction center. The endosymbiotic theory suggests that photosynthetic bacteria were acquired (by endocytosis) by early eukaryotic cells to form the first plant cells. Therefore, chloroplasts may be photosynthetic bacteria that adapted to life inside plant cells. Like mitochondria, chloroplasts possess their own DNA, separate from the nuclear DNA of their plant host cells and the genes in this chloroplast DNA resemble those found in cyanobacteria. DNA in chloroplasts codes for redox proteins such as those found in the photosynthetic reaction centers. The CoRR Hypothesis proposes that this Co-location is required for Redox Regulation.

Cyanobacteria and the Evolution of Photosynthesis

The biochemical capacity to use water as the source for electrons in photosynthesis evolved once, in a common ancestor of extant cyanobacteria. The geological record indicates that this transforming event took place early in Earth's history, at least 2450–2320 million years ago (Ma), and, it is speculated, much earlier. Because the Earth's atmosphere contained almost no oxygen during the estimated development of photosynthesis, it is believed that the first photosynthetic cyanobacteria did not generate oxygen. Available evidence from geobiological studies of Archean (>2500 Ma) sedimentary rocks indicates that life existed 3500 Ma, but the question of when oxygenic photosynthesis evolved is still unanswered. A clear paleontological window on cyanobacterial evolution opened about 2000 Ma, revealing an already-diverse biota of blue-green algae. Cyanobacteria remained the principal primary producers of oxygen throughout the Proterozoic Eon (2500–543 Ma), in part because the redox structure of the oceans favored photoautotrophs capable of nitrogen fixation. Green algae joined blue-green algae as the major primary producers of oxygen on continental shelves near the end of the Proterozoic, but it was only with the Mesozoic (251–65 Ma) radiations of dinoflagellates, coccolithophorids, and diatoms did the primary production of oxygen in marine shelf waters take modern form. Cyanobacteria remain critical to marine ecosystems as primary producers of oxygen in oceanic gyres, as agents of biological nitrogen fixation, and, in modified form, as the plastids of marine algae.

Discovery

Although some of the steps in photosynthesis are still not completely understood, the overall photosynthetic equation has been known since the 19th century.

Jan van Helmont began the research of the process in the mid-17th century when he carefully measured the mass of the soil used by a plant and the mass of the plant as it grew. After noticing that the soil mass changed very little, he hypothesized that the mass of the growing plant must come from the water, the only substance he added to the potted plant. His hypothesis was partially accurate — much of the gained mass also comes from carbon dioxide as well as water. However, this was a signaling point to the idea that the bulk of a plant's biomass comes from the inputs of photosynthesis, not the soil itself.

Joseph Priestley, a chemist and minister, discovered that, when he isolated a volume of air under an inverted jar, and burned a candle in it, the candle would burn out very quickly, much before it ran out of wax. He further discovered that a mouse could similarly "injure" air. He then showed that the air that had been "injured" by the candle and the mouse could be restored by a plant.

In 1778, Jan Ingenhousz, repeated Priestley's experiments. He discovered that it was the influence of sunlight on the plant that could cause it to revive a mouse in a matter of hours.

In 1796, Jean Senebier, a Swiss pastor, botanist, and naturalist, demonstrated that green plants consume carbon dioxide and release oxygen under the influence of light. Soon afterward, Nicolas-Théodore de Saussure showed that the increase in mass of the plant as it grows could not be due only to uptake of CO_2 but also to the incorporation of water. Thus, the basic reaction by which photosynthesis is used to produce food (such as glucose) was outlined.

Cornelis Van Niel made key discoveries explaining the chemistry of photosynthesis. By studying purple sulfur bacteria and green bacteria he was the first to demonstrate that photosynthesis is a light-dependent redox reaction, in which hydrogen reduces carbon dioxide.

Robert Emerson discovered two light reactions by testing plant productivity using different wavelengths of light. With the red alone, the light reactions were suppressed. When blue and red were combined, the output was much more substantial. Thus, there were two photosystems, one absorbing up to 600 nm wavelengths, the other up to 700 nm. The former is known as PSII, the latter is PSI. PSI contains only chlorophyll "a", PSII contains primarily chlorophyll "a" with most of the available chlorophyll "b", among other pigment. These include phycobilins, which are the red and blue pigments of red and blue algae respectively, and fucoxanthol for brown algae and diatoms. The process is most productive when the absorption of quanta are equal in both the PSII and PSI, assuring that input energy from the antenna complex is divided between the PSI and PSII system, which in turn powers the photochemistry.

Melvin Calvin works in his photosynthesis laboratory.

Robert Hill thought that a complex of reactions consisting of an intermediate to cytochrome b_6 (now a plastoquinone), another is from cytochrome f to a step in the carbohydrate-generating mechanisms. These are linked by plastoquinone, which does require energy to reduce cytochrome f for it is a sufficient reductant. Further experiments to prove that the oxygen developed during the photosynthesis of green plants came from water, were performed by Hill in 1937 and 1939. He showed that isolated chloroplasts give off oxygen in the presence of unnatural reducing agents like iron oxalate, ferricyanide or benzoquinone after exposure to light. The Hill reaction is as follows:

$$2 H_2O + 2 A + (\text{light, chloroplasts}) \rightarrow 2 AH_2 + O_2$$

where A is the electron acceptor. Therefore, in light, the electron acceptor is reduced and oxygen is evolved.

Samuel Ruben and Martin Kamen used radioactive isotopes to determine that the oxygen liberated in photosynthesis came from the water.

Melvin Calvin and Andrew Benson, along with James Bassham, elucidated the path of carbon assimilation (the photosynthetic carbon reduction cycle) in plants. The carbon reduction cycle is known as the Calvin cycle, which ignores the contribution of Bassham and Benson. Many scientists refer to the cycle as the Calvin-Benson Cycle, Benson-Calvin, and some even call it the Calvin-Benson-Bassham (or CBB) Cycle.

Nobel Prize-winning scientist Rudolph A. Marcus was able to discover the function and significance of the electron transport chain.

Otto Heinrich Warburg and Dean Burk discovered the I-quantum photosynthesis reaction that splits the CO_2, activated by the respiration.

Louis N.M. Duysens and Jan Amesz discovered that chlorophyll a will absorb one light, oxidize cytochrome f, chlorophyll a (and other pigments) will absorb another light, but will reduce this same oxidized cytochrome, stating the two light reactions are in series.

Development of the Concept

In 1893, Charles Reid Barnes proposed two terms, *photosyntax* and *photosynthesis*, for the biological process of *synthesis of complex carbon compounds out of carbonic acid, in the presence of chlorophyll, under the influence of light*. Over time, the term *photosynthesis* came into common usage as the term of choice. Later discovery of anoxygenic photosynthetic bacteria and photophosphorylation necessitated redefinition of the term.

C3 : C4 Photosynthesis Research

After WWII at late 1940 at the University of California, Berkeley, the details of photosynthetic carbon metabolism were sorted out by the chemists Melvin Calvin, Andrew Benson, James Bassham and a score of students and researchers utilizing the carbon-14 isotope and paper chromatography techniques. The pathway of CO_2 fixation by the algae *Chlorella* in a fraction of a second in light resulted in a 3 carbon molecule called phosphoglyceric acid (PGA). For that original and ground-breaking work, a Nobel Prize in Chemistry was awarded to Melvin Calvin 1961. In parallel, plant physiologists studied leaf gas exchanges using the new method of infrared gas analysis and a leaf chamber where the net photosynthetic rates ranged from 10 to 13 u mole CO_2/square metere.sec., with the conclusion that all terrestrial plants having the same photosynthetic capacities that were light saturated at less than 50% of sunlight. These rates were determined in potted plants grown indoors under low light intensity.

Later in 1958-1963 at Cornell University, field grown maize was reported to have much greater leaf photosynthetic rates of 40 u mol CO_2/square meter.sec and was not saturated at near full sunlight. This higher rate in maize was almost double those observed in other species such as wheat and soybean, indicating that large differences in photosynthesis exist among higher plants. At the University of Arizona, detailed gas exchange research on more than 15 species of monocot and dicot uncovered for the first time that differences in leaf anatomy are crucial factors in differentiating photosynthetic capacities among species. In tropical grasses, including maize, sorghum, sugarcane, Bermuda grass and in the dicot amaranthus, leaf photosynthetic rates were around 38–40 u mol CO_2/square meter.sec., and the leaves have two types of green cells, i. e. outer layer of mesophyll cells surrounding a tightly packed chlorophyllous vascular bundle sheath cells. This type of anatomy was termed Kranz anatomy in the 19th century by the botanist Gottlieb Haberlandt while studying leaf anatomy of sugarcane. Plant species with the greatest photosynthetic rates and Kranz anatomy showed no apparent photorespiration, very low CO_2 compensation point, high optimum temperature, high stomatal resistances and lower mesophyll resistances for gas diffusion and rates never saturated at full sun light. The research at Arizona was designated Citation Classic by the ISI 1986. These species was later termed C4 plants as the first stable compound of CO_2 fixation in light has 4 carbon as malate and aspartate. Other species that lack Kranz anatomy were termed C3 type such as cotton and sunflower, as the first stable carbon compound is the 3-carbon PGA acid. At 1000 ppm CO_2 in measuring air, both the C3 and C4 plants had similar leaf photosynthetic rates around 60 u mole CO_2/square meter. sec. indicating the suppression of phototorespiration in C3 plants.

Factors

The leaf is the primary site of photosynthesis in plants.

There are three main factors affecting photosynthesis and several corollary factors. The three main are:

- Light irradiance and wavelength
- Carbon dioxide concentration
- Temperature.

Light Intensity (Irradiance), Wavelength and Temperature

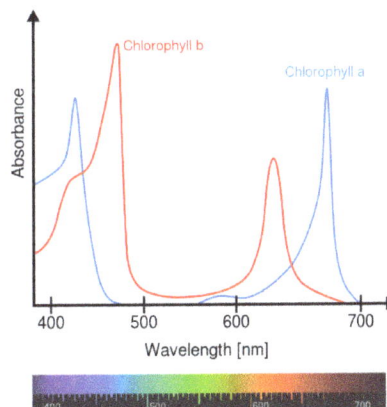

Absorbance spectra of free chlorophyll *a* (green) and *b* (red) in a solvent. The **action spectra** of chlorophyll molecules are slightly modified *in vivo* depending on specific pigment-protein interactions.

The process of photosynthesis provides the main input of free energy into the biosphere, and is one of four main ways in which radiation is important for plant life.

The radiation climate within plant communities is extremely variable, with both time and space.

In the early 20th century, Frederick Blackman and Gabrielle Matthaei investigated the effects of light intensity (irradiance) and temperature on the rate of carbon assimilation.

- At constant temperature, the rate of carbon assimilation varies with irradiance, increasing as the irradiance increases, but reaching a plateau at higher irradiance.

- At low irradiance, increasing the temperature has little influence on the rate of carbon assimilation. At constant high irradiance, the rate of carbon assimilation increases as the temperature is increased.

These two experiments illustrate several important points: First, it is known that, in general, photochemical reactions are not affected by temperature. However, these experiments clearly show that temperature affects the rate of carbon assimilation, so there must be two sets of reactions in the full process of carbon assimilation. These are, of course, the light-dependent 'photochemical' temperature-independent stage, and the light-independent, temperature-dependent stage. Second, Blackman's experiments illustrate the concept of limiting factors. Another limiting factor is the wavelength of light. Cyanobacteria, which reside several meters underwater, cannot receive the correct wavelengths required to cause photoinduced charge separation in conventional photosynthetic pigments. To combat this problem, a series of proteins with different pigments surround the reaction center. This unit is called a phycobilisome.

Carbon Dioxide Levels and Photorespiration

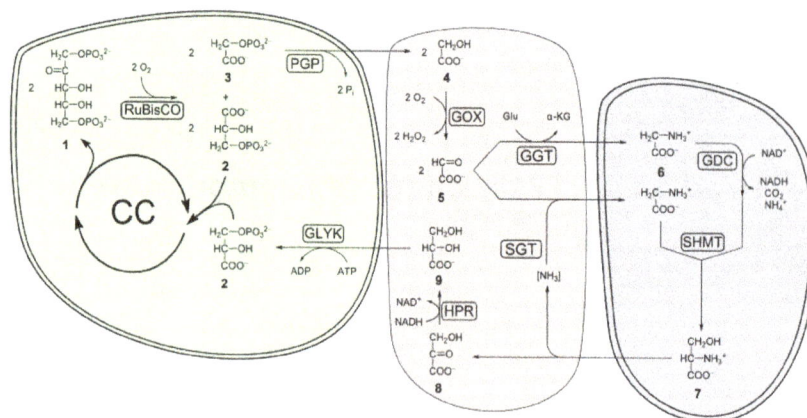

Photorespiration

As carbon dioxide concentrations rise, the rate at which sugars are made by the light-independent reactions increases until limited by other factors. RuBisCO, the enzyme that captures carbon dioxide in the light-independent reactions, has a binding affinity for both carbon dioxide and oxygen. When the concentration of carbon dioxide is high, RuBisCO will fix carbon dioxide. However, if the carbon dioxide concentration is low, RuBisCO will bind oxygen instead of carbon dioxide. This process, called photorespiration, uses energy, but does not produce sugars.

RuBisCO oxygenase activity is disadvantageous to plants for several reasons:

1. One product of oxygenase activity is phosphoglycolate (2 carbon) instead of 3-phosphoglycerate (3 carbon). Phosphoglycolate cannot be metabolized by the Calvin-Benson cycle and represents carbon lost from the cycle. A high oxygenase activity, therefore, drains the sugars that are required to recycle ribulose 5-bisphosphate and for the continuation of the Calvin-Benson cycle.

2. Phosphoglycolate is quickly metabolized to glycolate that is toxic to a plant at a high concentration; it inhibits photosynthesis.

3. Salvaging glycolate is an energetically expensive process that uses the glycolate pathway, and only 75% of the carbon is returned to the Calvin-Benson cycle as 3-phosphoglycerate. The reactions also produce ammonia (NH_3), which is able to diffuse out of the plant, leading to a loss of nitrogen.

A highly simplified summary is:

2 glycolate + ATP → 3-phosphoglycerate + carbon dioxide + ADP + NH_3

The salvaging pathway for the products of RuBisCO oxygenase activity is more commonly known as photorespiration, since it is characterized by light-dependent oxygen consumption and the release of carbon dioxide.

Phytochemistry

Phytochemistry is the study of phytochemicals, which are chemicals derived from plants. Those studying phytochemistry strive to describe the structures of the large number of secondary metabolic compounds found in plants, the functions of these compounds in human and plant biology, and the biosynthesis of these compounds. Plants synthesize phytochemicals for many reasons, including to protect themselves against insect attacks and plant diseases. Phytochemicals in food plants are often active in human biology, and in many cases have health benefits.

Phytochemistry can be considered sub-fields of botany or chemistry. Activities can be led in botanical gardens or in the wild with the aid of ethnobotany. The applications of the discipline can be for pharmacognosy, or the discovery of new drugs, or as an aid for plant physiology studies.

Techniques

Techniques commonly used in the field of phytochemistry are extraction, isolation, and structural elucidation (MS,1D and 2D NMR) of natural products, as well as various chromatography techniques (MPLC, HPLC, and LC-MS).

Constituent Elements

The list of simple elements of which plants are primarily constructed—carbon, oxygen, hydrogen, calcium, phosphorus, etc.—is not different from similar lists for animals, fungi, or even bacteria. The fundamental atomic components of plants are the same as for all life; only the details of the way in which they are assembled differs.

Eastern Medicine

Phytochemistry is widely used in the field of Chinese medicine especially in the field of herbal medicine.

Phytochemical technique mainly applies to the quality control of Chinese medicine, Ayurvedic medicine(Indian traditional medicine) or herbal medicine of various chemical components, such as saponins, alkaloids, volatile oils, flavonoids and anthraquinones. In the development of rapid and reproducible analytical techniques, the combination of HPLC with different detectors, such as diode array detector (DAD), refractive index detector (RID), evaporative light scattering detector (ELSD) and mass spectrometric detector (MSD), has been widely developed.

In most cases, biologically active compounds in Chinese medicine, Ayurveda, or herbal medicine have not been determined. Therefore, it is important to use the phytochemical methods to screen and analyze bioactive components, not only for the quality control of crude drugs, but also for the elucidation of their therapeutic mechanisms. Modern pharmacological studies indicate that binding to receptors or ion channels on cell membranes is the first step of some drug actions. A new method in phytochemistry called biochromatography has been developed. This method combines human red cell membrane extraction and high performance liquid chromatography to screen potential active components in Chinese medicine.

Types of Substances Studied

- Polyphenols

- Phytosterols

- Alkaloids

- Saponins

Major Research Institutes

- Tropical Botanical Garden and Research Institute

- UBC Botanical Garden and Centre for Plant Research

Seedling

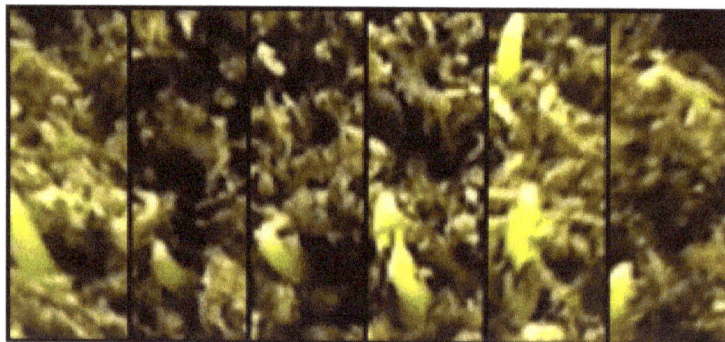

Grass seedlings, 150-minute time lapse

A seedling is a young plant sporophyte developing out of a plant embryo from a seed. Seedling development starts with germination of the seed. A typical young seedling consists of three main parts: the radicle (embryonic root), the hypocotyl (embryonic shoot), and the cotyledons (seed leaves). The two classes of flowering plants (angiosperms) are distinguished by their numbers of seed leaves: monocotyledons (monocots) have one blade-shaped cotyledon, whereas dicotyledons (dicots) possess two round cotyledons. Gymnosperms are more varied. For example, pine seedlings have up to eight cotyledons. The seedlings of some flowering plants have no cotyledons at all. These are said to be acotyledons.

The plumule is the part of a seed embryo that develops into the shoot bearing the first true leaves of a plant. In most seeds, for example the sunflower, the plumule is a small conical structure without any leaf structure. Growth of the plumule does not occur until the cotyledons have grown above ground. This is epigeal germination. However, in seeds such as the broad bean, a leaf structure is visible on the plumule in the seed. These seeds develop by the plumule growing up through the soil with the cotyledons remaining below the surface. This is known as hypogeal germination.

Photomorphogenesis and Etiolation

Dicot seedlings grown in the light develop short hypocotyls and open cotyledons exposing the epicotyl. This is also referred to as photomorphogenesis. In contrast, seedlings grown in the dark develop long hypocotyls and their cotyledons remain closed around the epicotyl in an *apical hook*. This is referred to as skotomorphogenesis or etiolation. Etiolated seedlings are yellowish in color as chlorophyll synthesis and chloroplast development depend on light. They will open their cotyledons and turn green when treated with light.

In a natural situation, seedling development starts with skotomorphogenesis while the seedling is growing through the soil and attempting to reach the light as fast as possible. During this phase, the cotyledons are tightly closed and form the *apical hook* to protect the shoot apical meristem from damage while pushing through the soil. In many plants, the seed coat still covers the cotyledons for extra protection.

Upon breaking the surface and reaching the light, the seedling's developmental program is switched to photomorphogenesis. The cotyledons open upon contact with light (splitting the seed coat open, if still present) and become green, forming the first photosynthetic organs of the young plant. Until this stage, the seedling lives off the energy reserves stored in the seed. The opening of the cotyledons exposes the shoot apical meristem and the *plumule* consisting of the first *true leaves* of the young plant.

The seedlings sense light through the light receptors phytochrome (red and far-red light) and cryptochrome (blue light). Mutations in these photo receptors and their signal transduction components lead to seedling development that is at odds with light conditions, for example seedlings that show photomorphogenesis when grown in the dark.

Seedling Growth and Maturation

Once the seedling starts to photosynthesize, it is no longer dependent on the seed's energy reserves. The apical meristems start growing and give rise to the root and shoot. The first "true" leaves expand and can often be distinguished from the round cotyledons through their species-dependent distinct shapes. While the plant is growing and developing additional leaves, the cotyledons eventually senesce and fall off. Seedling growth is also affected by mechanical stimulation, such as by wind or other forms of physical contact, through a process called thigmomorphogenesis.

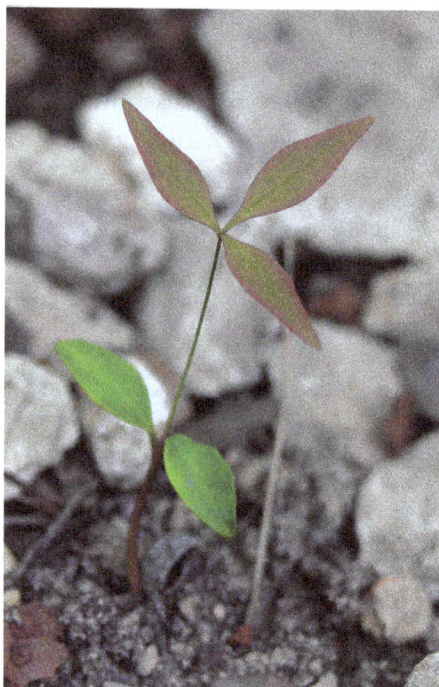

Seedling of a dicot, *Nandina domestica*, showing two green cotyledon leaves, and the first "true" leaf with its distinct leaflets and red-green color.

Temperature and light intensity interact as they affect seedling growth; at low light levels about 40 lumens/m² a day/night temperature regime of 28 °C/13 °C is effective (Brix 1972). A photoperiod shorter than 14 hours causes growth to stop, whereas a photoperiod extended with low light intensities to 16 h or more brings about continuous (free) growth. Little is gained by using more than 16 h of low light intensity once seedlings are in the free growth mode. Long photoperiods using high light intensities from 10,000 to 20,000 lumens/m² increase dry matter production, and increasing the photoperiod from 15 to 24 hours may double dry matter growth (Pollard and Logan 1976, Carlson 1979).

The effects of carbon dioxide enrichment and nitrogen supply on the growth of white spruce and trembling aspen were investigated by Brown and Higginbotham (1986). Seedlings were grown in controlled environments with ambient or enriched atmospheric CO_2 (350 or 750 fl/L, respectively) and with nutrient solutions with high, medium, and low N content (15.5, 1.55, and 0.16 mM). Seedlings were harvested, weighed, and measured at intervals of less than 100 days. N supply strongly affected biomass accumulation, height, and leaf area of both species. In white spruce only, the root weight ratio (RWR) was significantly increased with the low-nitrogen regime. CO_2 enrichment for 100 days significantly increased the leaf and total biomass of white spruce seedlings in the high-N regime, RWR of seedlings in the medium-N regime, and root biomass of seedlings in the low-N regime.

First-year seedlings typically have high mortality rates, drought being the principal cause, with roots having been unable to develop enough to maintain contact with soil sufficiently moist to prevent the development of lethal seedling water stress. Somewhat paradoxically, however, Eis (1967a) observed that on both mineral and litter seedbeds, seedling mortality was greater in moist habitats (alluvium and *Aralia–Dryopteris*) than in dry habitats (*Cornus*–Moss). He commented that in dry habitats after the first growing season surviving seedlings appeared to have a much better chance of continued survival than those in moist or wet habitats, in which frost heave and competition from lesser vegetation became major factors in later years. The annual mortality documented by Eis (1967a) is instructive.

Pests and Diseases

Seedlings are particularly vulnerable to attack by pests and diseases and can consequently experience high mortality rates. Pests and diseases which are especially damaging to seedlings include damping off, cutworms, slugs and snails.

Transplanting

Seedlings are generally transplanted when the first pair of true leaves appear. A shade may be provided if the area is arid or hot. A commercially available vitamin hormone concentrate may be used to avoid transplant shock which may contain thiamine hydrochloride, naphthly acetic acid and indole butyric acid.

Flora

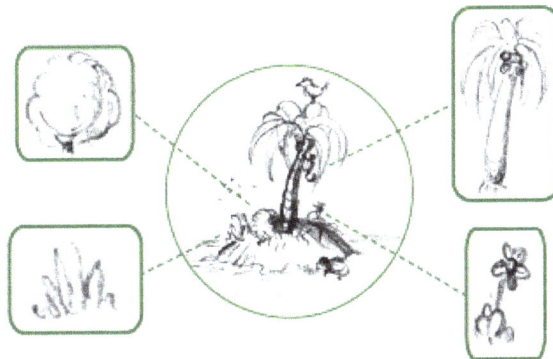

Simplified schematic of an island's flora - all its plant species, highlighted in boxes.

Flora is the plant life occurring in a particular region or time, generally the naturally occurring or indigenous—native plant life. The corresponding term for animal life is fauna. *Flora, fauna* and other forms of life such as fungi are collectively referred to as biota. Sometimes bacteria and fungi are also referred to as flora, as in the terms gut flora or skin flora.

Etymology

The word "flora" comes from the Latin name of Flora, the goddess of plants, flowers, and fertility in Roman mythology.

The distinction between vegetation (the general appearance of a community) and flora (the taxonomic composition of a community) was first made by Jules Thurmann (1849). Prior to this, the two terms were used indiscriminately.

Flora Classifications

Plants are grouped into floras based on region (floristic regions), period, special environment, or climate. Regions can be geographically distinct habitats like mountain vs. flatland. Floras can mean plant life of a historic era as in *fossil flora*. Lastly, floras may be subdivided by special environments:

- *Native flora*. The native and indigenous flora of an area.

- *Agricultural and horticultural flora (garden flora)*. The plants that are deliberately grown by humans.

- *Weed flora*. Traditionally this classification was applied to plants regarded as undesirable, and studied in efforts to control or eradicate them. Today the designation is less often used as a classification of plant life, since it includes three different types of plants: weedy species, invasive species (that may or may not be weedy), and native and introduced non-weedy species that are agriculturally undesirable. Many native plants previously considered weeds have been shown to be beneficial or even necessary to various ecosystems.

Documentation of Floras

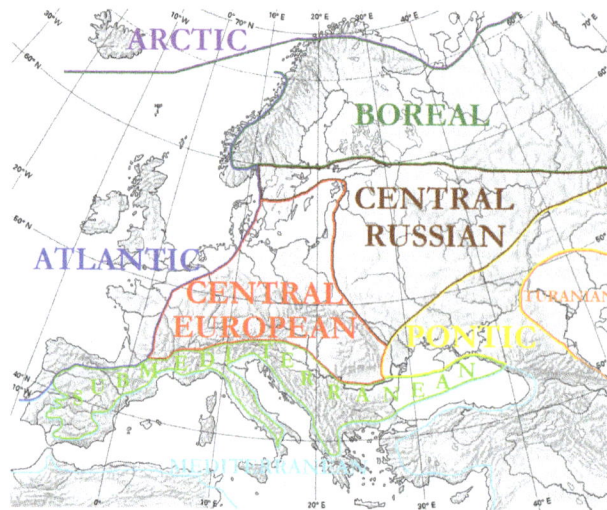

Floristic regions in Europe according to Wolfgang Frey and Rainer Lösch

The flora of a particular area or time period can be documented in a publication also known as a "flora" (often capitalized as "Flora" to distinguish the two meanings when they might be confused). Floras may require specialist botanical knowledge to use with any effectiveness. Traditionally they are books, but some are now published on CD-ROM or websites.

It is said that the *Flora Sinensis* by the Polish Jesuit Michał Boym was the first book that used the name "Flora" in this meaning, a book covering the plant world of a region. However, despite its title it covered not only plants, but also some animals of the region.

A published flora often contains diagnostic keys. Often these are *dichotomous* keys, which require the user to repeatedly examine a plant, and decide which one of two alternatives given best applies to the plant.

Tropism

Phycomyces, a fungus, exhibiting phototropism

A tropism (from Greek *tropos*, "a turning") is a biological phenomenon, indicating growth or turning movement of a biological organism, usually a plant, in response to an environmental stimulus. In tropisms, this response is dependent on the direction of the stimulus (as opposed to nastic movements which are non-directional responses). Viruses and other pathogens also affect what is called "host tropism", "tissue tropism", or "cell tropism", or in which case tropism refers to the way in which different viruses/pathogens have evolved to preferentially target specific host species, specific tissue, or specific cell types within those species. Tropisms are usually named for the stimulus involved (for example, a phototropism is a reaction to sunlight) and may be either *positive* (towards the stimulus) or *negative* (away from the stimulus).

Tropisms occur in four sequential steps. First, there is a perception to a stimulus, which is usually beneficiary to the plant. Next, signal transduction occurs. This leads to auxin redistribution at the cellular level and finally, the growth response occurs.

Tropisms are typically associated with plants (although not necessarily restricted to them). Where an organism is capable of directed physical movement (motility), movement or activity in response to a specific stimulus is more likely to be regarded by behaviorists as a *taxis* (directional response) or a *kinesis* (non-directional response).

In English, the word *tropism* is used to indicate an action done without cognitive thought: However, "tropism" in this sense has a proper, although non-scientific, meaning as an innate tendency, natural inclination, or propensity to act in a certain manner towards a certain stimulus.

In botany, the Cholodny–Went model, proposed in 1927, is an early model describing tropism in emerging shoots of monocotyledons, including the tendencies for the stalk to grow towards light (phototropism) and the roots to grow downward (gravitropism). In both cases the directional growth is considered to be due to asymmetrical distribution of auxin, a plant growth hormone.

Types

In plants (and bacteria)

Example of gravitropism in the remains of a cellar of a Roman villa in the
Archeologic Park in Baia, Italy

- Aerotropism, growth of plants towards or away from a source of oxygen

- Chemotropism, movement or growth in response to chemicals

- Electrotropism, movement or growth in response to an electric field

- Exotropism, continuation of growth "outward," i.e. in the previously established direction

- Geotropism (or gravitropism), movement or growth in response to gravity

  - Apogeotropism, negative geotropism

- Heliotropism, diurnal motion or seasonal motion of plant parts in response to the direction of the sun, (e.g. the sunflower)

  - Apheliotropism, negative heliotropism

- Hydrotropism, movement or growth in response to water. In plants, the root cap senses differences in water moisture in the soil, and signals cellular changes that causes the root to curve towards the area of higher moisture.

  - Prohydrotropism, positive hydrotropism

- Hygrotropism, movement or growth in response to moisture or humidity

- Magnetotropism, movement or growth in response to magnetic fields

- Orthotropism, movement or growth in the same line of action as the stimulus.

- Plagiotropism, movement or growth at an angle to a line of stimulus such as gravity or light.

- Phototropism, movement or growth in response to lights or colors of light

  o   Aphototropism, negative phototropism

  o   Skototropism, negative phototropism of vines

- Thermotropism, movement of growth in response to temperature

- Thigmotropism, movement or growth in response to touch or contact

In Viruses

- Amphotropism, wide host range (e.g. infects many species or cell types)

- Ecotropism, limited host range (e.g. infects only one species or cell type)

- HIV tropism, the means of entry into cells used by a given strain of HIV

- Neurotropism, a virus that preferentially infects the host's nervous system.

References

- Mullineaux CW (1999). "The thylakoid membranes of cyanobacteria: structure, dynamics and function". Australian Journal of Plant Physiology. 26 (7): 671–677. doi:10.1071/PP99027

- Raven PH, Evert RF, Eichhorn SE (2005). Biology of Plants, (7th ed.). New York: W. H. Freeman and Company. pp. 124–127. ISBN 0-7167-1007-2

- Olson JM (May 2006). "Photosynthesis in the Archean era". Photosynthesis Research. 88 (2): 109–17. PMID 16453059. doi:10.1007/s11120-006-9040-5

- Maxwell K, Johnson GN (Apr 2000). "Chlorophyll fluorescence--a practical guide". Journal of Experimental Botany. 51 (345): 659–68. PMID 10938857. doi:10.1093/jexbot/51.345.659

- Monson RK, Sage RF (1999). "The Taxonomic Distribution of C$_4$ Photosynthesis". C$_4$ plant biology. Boston: Academic Press. pp. 551–580. ISBN 0-12-614440-0

- Miyamoto K. "Chapter 1 – Biological energy production". Renewable biological systems for alternative sustainable energy production (FAO Agricultural Services Bulletin – 128). Food and Agriculture Organization of the United Nations. Retrieved 2009-01-04

- Venn AA, Loram JE, Douglas AE (2008). "Photosynthetic symbioses in animals". Journal of Experimental Botany. 59 (5): 1069–80. PMID 18267943. doi:10.1093/jxb/erm328

- Gale, Joseph (2009). Astrobiology of Earth : The emergence, evolution and future of life on a planet in turmoil. OUP Oxford. pp. 112–113. ISBN 9780191548352

- Bryant DA, Frigaard NU (Nov 2006). "Prokaryotic photosynthesis and phototrophy illuminated". Trends in Microbiology. 14 (11): 488–96. PMID 16997562. doi:10.1016/j.tim.2006.09.001

- Herrero A, Flores E (2008). The Cyanobacteria: Molecular Biology, Genomics and Evolution (1st ed.). Caister Academic Press. ISBN 978-1-904455-15-8

- Tavano CL, Donohue TJ (Dec 2006). "Development of the bacterial photosynthetic apparatus". Current Opinion in Microbiology. 9 (6): 625–31. PMC 2765710. PMID 17055774. doi:10.1016/j.mib.2006.10.005

- John T. Arnason; Rachel Mata; John T. Romeo (2013-11-11). "Phytochemistry of Medicinal Plants". Springer Science & Business Media. ISBN 9781489917782

- "World Consumption of Primary Energy by Energy Type and Selected Country Groups, 1980–2004". Energy Information Administration. July 31, 2006. Archived from the original (XLS) on November 9, 2006. Retrieved 2007-01-20

- Douglas SE (Dec 1998). "Plastid evolution: origins, diversity, trends". Current Opinion in Genetics & Development. 8 (6): 655–61. PMID 9914199. doi:10.1016/S0959-437X(98)80033-6

- Dodd AN, Borland AM, Haslam RP, Griffiths H, Maxwell K (Apr 2002). "Crassulacean acid metabolism: plastic, fantastic". Journal of Experimental Botany. 53 (369): 569–80. PMID 11886877. doi:10.1093/jexbot/53.369.569

Permissions

All chapters in this book are published with permission under the Creative Commons Attribution Share Alike License or equivalent. Every chapter published in this book has been scrutinized by our experts. Their significance has been extensively debated. The topics covered herein carry significant information for a comprehensive understanding. They may even be implemented as practical applications or may be referred to as a beginning point for further studies.

We would like to thank the editorial team for lending their expertise to make the book truly unique. They have played a crucial role in the development of this book. Without their invaluable contributions this book wouldn't have been possible. They have made vital efforts to compile up to date information on the varied aspects of this subject to make this book a valuable addition to the collection of many professionals and students.

This book was conceptualized with the vision of imparting up-to-date and integrated information in this field. To ensure the same, a matchless editorial board was set up. Every individual on the board went through rigorous rounds of assessment to prove their worth. After which they invested a large part of their time researching and compiling the most relevant data for our readers.

The editorial board has been involved in producing this book since its inception. They have spent rigorous hours researching and exploring the diverse topics which have resulted in the successful publishing of this book. They have passed on their knowledge of decades through this book. To expedite this challenging task, the publisher supported the team at every step. A small team of assistant editors was also appointed to further simplify the editing procedure and attain best results for the readers.

Apart from the editorial board, the designing team has also invested a significant amount of their time in understanding the subject and creating the most relevant covers. They scrutinized every image to scout for the most suitable representation of the subject and create an appropriate cover for the book.

The publishing team has been an ardent support to the editorial, designing and production team. Their endless efforts to recruit the best for this project, has resulted in the accomplishment of this book. They are a veteran in the field of academics and their pool of knowledge is as vast as their experience in printing. Their expertise and guidance has proved useful at every step. Their uncompromising quality standards have made this book an exceptional effort. Their encouragement from time to time has been an inspiration for everyone.

The publisher and the editorial board hope that this book will prove to be a valuable piece of knowledge for students, practitioners and scholars across the globe.

Index

www.ingramcontent.com/pod-product-compliance
Lightning Source LLC
Chambersburg PA
CBHW082031190326
41458CB00010B/3329